BIRDS OF CUBA

A PHOTOGRAPHIC GUIDE

Arturo Kirkconnell, Patricia E. Bradley and
Yves-Jacques Rey-Millet

Comstock Publishing Associates
an imprint of
Cornell University Press
Ithaca, New York

Copyright © Arturo Kirkconnell, Patricia E. Bradley and Yves-Jacques Rey-Millet, 2020
Photographs © as credited on page 377, 2020
For legal purposes, the Acknowledgements on page 9 constitute an extension of this
copyright page

All rights reserved. Except for brief quotations in a review, this book, or parts
thereof, must not be reproduced in any form without permission in writing from the
publisher. For information, address Cornell University Press, Sage House, 512 East
State Street, Ithaca, New York 14850. Visit our website at cornellpress.cornell.edu.

Published in the United States of America in 2020 by Cornell University Press
First Published in Great Britain in 2020 by Bloomsbury Publishing Plc

Library of Congress Cataloging-in-Publication Data

Names: Kirkconnell, Arturo, author. | Bradley, Patricia (Patricia E.), author. |
    Rey-Millet, Yves-Jacques, author, photographer.
Title: Birds of Cuba : a photographic guide / Arturo Kirkconnell, Patricia E. Bradley,
    Yves-Jacques Rey-Millet.
Description: Ithaca [New York] : Cornell University Press, 2020. |
    Includes bibliographical references and index.
Identifiers: LCCN 2020011451 (print) | LCCN 2020011452 (ebook) |
    ISBN 9781501751561 (paperback) | ISBN 9781501751578 (epub)
Subjects: LCSH: Birds—Cuba—Geographical distribution. |
    Birds—Cuba—Identification. | Birds—Cuba—Pictorial works.
Classification: LCC QL688.C9 K57 2020 (print) | LCC QL688.C9 (ebook) |
    DDC 598.097291—dc23
LC record available at https://lccn.loc.gov/2020011451
LC ebook record available at https://lccn.loc.gov/2020011452

Maps by Julian Baker
Design by Rod Teasdale

Printed and bound in India by Replika Press Pvt. Ltd.

For Yves-Jacques Rey-Millet (1946–2016)

A lifetime of photographing birds
throughout the West Indies
has contributed to our understanding
of this precious resource.

AK: To my wonderful grandchildren
Adam and Alexa Kirkconnell,
whom I love.

PEB: To my beloved grandson
Alexander Michael Bradley.

CONTENTS

- PREFACE 8
- ACKNOWLEDGEMENTS 9

INTRODUCTION
- Geographical position 10
- About Cuba 11
- Climate 11
- Geology 12
- Origins of the avifauna 12
- Vegetation and habitats 13
- History of ornithology in Cuba 23
- The avifauna 24
- Conservation 31
- Where to watch birds on Cuba 33
- Information for visiting birders 38
- How to use this guide 39
- Bird topography 41

SPECIES ACCOUNTS
- Grebes 43
- Petrels 44
- Shearwaters 44
- Tropicbirds 45
- Boobies 46
- Pelicans 47
- Cormorants 49
- Anhinga 51
- Frigatebirds 52
- Herons and Egrets 53
- Ibises 65
- Spoonbill 67
- Wood Storks 68
- Flamingos 69
- Ducks 70
- Osprey 89
- New World Vultures 90
- Hawks and Eagles 91
- Falcons 103
- Rails 107
- Limpkin 117
- Sandhill Crane 118
- Plovers 119
- Oystercatchers 126
- Stilts and Avocets 127
- Jacana 129
- Sandpipers and allies 130
- Jaegers 149
- Gulls 150
- Terns 156
- Skimmers 168
- Pigeons and Doves 169
- Parrots and Parakeets 183
- Cuckoos 186
- Barn Owl 191
- Typical Owls 192
- Nightjars 199
- Potoos 203
- Swifts 204
- Hummingbirds 207
- Trogons 211
- Todies 213
- Kingfishers 215
- Woodpeckers 216
- Tyrant Flycatchers 223
- Vireos 236
- Crows 244
- Swallows 247
- Wrens 254
- Kinglets 256
- Gnatcatchers 257
- Solitaires 260
- Thrushes 262
- Mockingbirds 268
- Waxwings 270
- New World Warblers 271
- Bananaquit 306
- Cardinals 307
- Spindalis 309
- Tanagers 310
- Bullfinches, Grassquits and Sparrows 311
- Grosbeaks and Buntings 323
- New World Blackbirds and allies 329

APPENDICES
- A Vagrants recorded on the Cuban archipelago 342
- B Introduced species breeding on Cuba 367
- C Species likely to be extinct on the Cuban archipelago 371
- D Endemic Species 372
- E Endemic Subspecies 374

- BIBLIOGRAPHY 376
- IMAGE CREDITS 377
- INDEX OF ENGLISH NAMES 378
- INDEX OF SCIENTIFIC NAMES 381

PREFACE

The preparation of this guide has not been easy due to the death in June 2016 of the third author and principal photographer, Yves-Jacques Rey-Millet, after a brief but devastating illness. It was a terrible shock and work on the book ground to a halt. His ambition had been to complete a trio of West Indian photographic field guides: Cayman Islands, Jamaica and Cuba. Yves-Jacques and Patricia had worked together on many projects since 1983 and the loss of his friendship and professional collaboration has been profound. Arturo, too, had developed a special bond with Yves-Jacques since they began working on the Cuban book in 2013. In late 2017, we returned to completing the text and the daunting task of reviewing over 400,000 of Yves-Jacques's unsorted photographs. Despite this vast body of work, he had not had sufficient time to complete the Cuban portfolio.

This is the first photographic field guide for the birds of Cuba, with a total of 384 recorded species, of which 29 are endemic species. Among these are outstanding birds such as the Bee Hummingbird (the smallest bird in the world), the charming and enigmatic Blue-headed Quail-Dove, the Cuban Trogon with its unique tail, the very colourful Cuban Tody and two beautiful picids, Cuban Green Woodpecker and Fernandina's Flicker. A total of 156 species breed on Cuba, 138 Nearctic migrants are regular winter visitors, 44 species are exclusively passage migrants, 74 are currently considered to be vagrants or accidental visitors, and 16 are summer residents that come to Cuba to breed and spend our winters in the south. Eight well-established species are confirmed as introduced and another two were possibly introduced. The entries for three species use illustrations of birds for which we were unable to obtain photographs because of their rarity.

Arturo Kirkconnell and Patricia E. Bradley
Cuba and the Cayman Islands
August 2020

Blue-headed Quail-Dove.

ACKNOWLEDGMENTS

We are especially grateful to those who have come to our rescue by sharing their images to fill the empty spaces; in particular, we would like to thank Nancy Norman, Arturo Kirkconnell Jr and Bruce Hallett. And for the task of preparing the digital images for publication we thank the extraordinary and extensive work of Esteban Gutiérrez, Nancy Norman and Arturo Kirkconnell Jr. We could not have completed the field guide without the generosity of these four friends giving their time and expertise. Thanks to Glen Tepke and Yves Aubry for sharing last-minute images. We would like to thank our Publisher, Jim Martin, for his patience and understanding throughout this project, our Editor, Jenny Campbell, for her patience and wise judgement in producing the book amid our deluge of changes, and to Rod Teasdale for his design skills.

We also wish to thank Alexandra Günther-Calhoun for giving permission to use the Rey-Millet photographs; Eldon and Pat Kirkconnell for their support during Arturo's visits to the Cayman Islands; Jim Wiley for reviewing parts of the Introduction; and artist Alvaro de Jesús for his wonderful illustrations. We would also like to thank Carlos Mancina who kindly cooperated with the production of very accurate maps; and Ramona Oviedo, Manuel Iturralde-Vinent and Jesús Pajón who provided important information for the Introduction.

Thanks, too, to KBC and Alvaro's Adventures for supporting some of the field expeditions providing data and images; the American Museum of Natural History, New York, for permission to review bird collections, with special thanks to Paul Sweet; Dan Thompson at Vireo Resources for Ornithology, Drexel University; Rosita M. Posada for all her support, and Roberto Posada, who assisted with some of the field expeditions; and Marlene Concepción for facilitating communication.

Female Fernandina's Flicker.

INTRODUCTION TO THE CUBAN ARCHIPELAGO

GEOGRAPHICAL POSITION

Cuba is the largest country in the Caribbean Sea and represents more than half of the total land mass of the West Indies. It is bordered to the north by the Straits of Florida, to the north-east by the Nicholas Channel and the Old Bahama Channel, to the south by the Windward Passage and the Cayman Trench, while to the south-west lies the Caribbean Sea. To the west, it reaches to the Yucatán Channel and the north-west is open to the Gulf of Mexico.

The Cuban archipelago comprises a main island (Cuba), the Isle of Pines (or Isle of Youth) and more than 3,000 small islands and cays (generally small, low-elevation islands on the surface of reefs) comprising four archipelagos. Off the southern coast lie the archipelagos of Jardines de la Reina and Los Canarreos. Off the north-eastern shore lies the Sabana-Camagüey archipelago or Jardines del Rey, which is composed of approximately 2,517 cays and islands, including Cayo Coco, Cayo Romano and Cayo Paredón Grande. The Colorados archipelago lies off the north-western coast.

Map of the Cuban archipelago.

Geographical statistics of the Cuban archipelago:
Main island: 104,553km².
Isle of Pines (or Isle of Youth): 2,204km².
Cays: 3,126km².
Total area: 109,883km².

Geographical limits:
Northern limit: Cayo Cruz del Padre, 23° 17′ 09″ N.
Southern limit: Punta del Inglés, 19° 49′ 36″ N.
Eastern limit: Punta del Quemado, 74° 07′ 52″ E.
Western limit: Cabo de San Antonio, 84° 57′ 54″ W.

Neighbouring countries:
Haiti: 77km to the east, across the Windward Passage.
Jamaica: 140km to the south, across the Strait of Columbus.
United States: 180km to the north, across the Straits of Florida.
Mexico: 210km to the west, across the Yucatán Channel.

Length and breadth of the main island:
Longitudinal axis: 1,250km.
Widest part: 191km.
Narrowest: 31km.

Highest elevations:
Western provinces: Pan de Guajaibón; 699m, Cordillera de Guaniguanico.
Central provinces: Pico de San Juan, 1,140m, Massif of Guamuhaya.
Eastern provinces: Pico Turquino, 1,972m, in the Sierra Maestra.

Length of the main rivers:
Flowing north: Sagua la Grande, 144km; Caonao, 132km; Toa, 118km.
Flowing south: Cauto, 343km; Zaza, 145km; Agabama, 118km.

ABOUT CUBA

The total Cuban population was c. 11.5 million in 2017. The first Spanish settlement, in Baracoa, was founded by Diego Velázquez in 1511. There were two major population waves from Spain, of c. 0.5 million in 1820–98, and 1.2 million in 1899–1930, although many returned. During 1842–73, 221,000 Africans were brought to Cuba as slaves. In 1959, a revolutionary government took power, and by 1961 a socialist system had been established. The present constitution ascribes the role of the Communist Party of Cuba to be the 'leading force of the society and the state' and as such has the capability of setting national policy. The language spoken is Spanish.

CLIMATE

The Cuban climate is tropical but varies considerably, due to the geographical location of the island and seasonal variations. The dry period extends generally from November–April. The driest regions are on the south coast: the south-eastern province of Guantánamo and east of Cienfuegos, where annual rainfall is less than 200mm. The annual mean temperature on the island is 25.2°C, mean humidity 80 per cent and mean annual rainfall 1,374mm. The Sierra Maestra mountain range receives an average of 1,600mm. The maximum recorded annual rainfall is in the eastern mountain range (Nipe–Sagua–Baracoa), with mean rainfall exceeding 3,400mm. Wind direction varies with locality. Along the north coast and inland, prevailing winds are from the north and the east-north-east. Along the south coast, winds are mostly from the north-east to south-east.

The winter is marked by the arrival of cold fronts from the north. These usually affect the western two-thirds of the archipelago and occur from September–March. From February–April, southerly winds predominate.

The Caribbean experiences a strong tropical storm season from June–November. In the last 168 years, Cuba has averaged slightly more than one tropical storm per year, with approximately 196 storms crossing the territory in that period. Of these, 37 were classified as high intensity, or hurricanes, with winds reaching more than 210km per hour. Storms are more frequent in the western third of the island.

Annual precipitation across the Cuban archipelago.

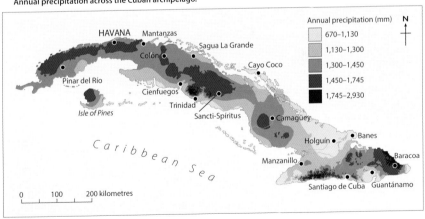

GEOLOGY

Origin of the Cuban land mass

The islands of the Greater Antilles (Cuba, Hispaniola, Puerto Rico and Jamaica) have complex geological histories, and the geological foundation of Cuba is an amalgam of rocks of various origins. The general trend on Cuba's tectonic evolution has been the rising of the land surface and an increase in its area; this process commenced *c.* 40–45 million years ago (MYA). Around 3–4 MYA, the Cuban archipelago was part of a large promontory in the north-western Caribbean. From about 20,000 years ago to the present, glacial and interglacial periods alternated, with corresponding falls in sea level (creating more land) followed by sea level rises (which flooded many land areas). It is estimated that the average speed of uplift of the Cuban land mass during this period ranged from 1–3mm per year, while the mean rise in sea level was 4.8 mm. Since *c.* 8,000–10,000 years ago, sea level rise first accelerated, then subsequently declined, redrawing the coastline to some extent and adjusting the position of inshore coral reefs.

ORIGINS OF THE AVIFAUNA

The oldest bird fossil recorded in the West Indies is a feather found preserved in amber on Hispaniola and estimated to date from 15–20 MYA; it belonged to a woodpecker closely related to Antillean Piculet. The oldest mammal fossils found on Cuba (a rodent, a sloth and a monkey) date from 20–30 MYA. If mammals, which are much less able to disperse across physical barriers than birds, were present in the Greater Antilles, it is logical that an indigenous avifauna was present, too.

During the late Oligocene (*c.* 25 MYA), the eastern part of Cuba was connected to Hispaniola and this connection lasted *c.* 10 million years. Later, *c.* 16–14 MYA, tectonic activity led to the separation of land blocks to form the Greater Antillean islands of today; Cuba and Hispaniola separated, and the Paso de los Vientos formed. Since the separation of the Greater Antilles, faunal fragmentation, or vicariance, among islands has given a new biogeographical panorama to the region, as evidenced by fossil remains shared among islands and by the present distribution of some species, including Cuban and Hispaniolan Trogons, and the nightjar genus *Siphonorhis*.

During the Pliocene (2.6–5.0 MYA), birds reached their maximum diversity. New genera appeared, and most of those present-day species evolved. In the early Pliocene, sea level was *c.* 80m higher than it is at present and most of Cuba was submerged. Yet, there is no doubt that birds were part of the terrestrial fauna, and would have found refuge in the highest mountains, as they did in Jamaica, Hispaniola and Puerto Rico. At some point during the last 3–4 million years, the three islands were reunited to form a single island. The faunas of the reunited islands dispersed throughout Cuba although during interglacial periods, isolation and fragmentation of populations continued. Such vicariance could also probably explain the modern distribution of the genus *Teretistris*.

In the mid- to late Pliocene (2.6–3.0 MYA), a pattern of alternating glacial and interglacial cycles commenced and continued during the Pleistocene (2.6 MYA to 11,000 years ago). During glacial stages, a cold, dry climate with arid conditions predominated in the Greater Antilles, while during interglacial periods, a warm, humid climate was dominant. Rising and falling sea level also affected ecosystems and habitats throughout the region and consequently modified the flora and fauna.

Avian colonisers

The ancestors of the present Greater Antilles endemic avifauna have spread from west to east and from north to south. The most accepted biogeographic theory is that over-water dispersal from Middle and North America was responsible for the ancestral taxa of present Greater Antillean avifauna. Glaciation and falling sea levels caused island land masses to increase in area, reducing the distances between the islands and the continent. Thus, many species colonised the Caribbean from Middle America via Jamaica (believed to be the first Greater Antillean island to be colonised) and via Yucatán to western Cuba. During Pleistocene glaciations (*c.* 17,000–20,000 years ago), sea level was 100–120m below the present level, resulting in Caribbean islands being larger than at present. For example, Puerto Rico and the Virgin Islands were connected, and many of the Bahamian islands were united. When sea levels were 40m below present levels, c. 10,000–12,000 years ago, the Great Bahama Bank was only 20km from Cuba (currently c. 180km separates Cuba from the Great Bahama Bank) and had a land area of 102,231–103,670km^2, nearly the same size as Cuba today (109,833km^2). Consequently, several taxa, such as Northern Mockingbird and Northern Flicker, entered the West Indies from North America via Florida to Cuba and the Bahamas.

Over-water dispersal from the Middle and North America continent was likely in response to global climate changes in tropical and temperate regions, and to the availability of food resources in the breeding and wintering grounds. It acted as a selective filter, preventing the colonisation of distant sites by groups of birds with weak flight capabilities and allowing taxa with advantageous ecological characteristics to colonise successfully. Ancestral migratory movements throughout the Greater Antilles were probably also achieved via regular migrations or by post-breeding dispersal during these glacial epochs. The existence of natural corridors in or near the distribution range of a particular species may facilitate its dispersal. For example, since the mid-Pleistocene, a corridor of thorn-scrub habitat along the North American Gulf coast has facilitated interchange between eastern Mexico and Texas and south-east to the Florida Peninsula; it persists today.

Endemism
During glacial and interglacial events earlier in the Pleistocene, the distribution of West Indian vertebrates was affected by alternating emergence and submergence of land. This cycle resulted in repeated faunal isolation, speciation and extinction. Once resident populations were established, endemic taxa evolved over time by a process that probably accelerated due to greater isolation during interglacial periods, when sea level was higher and there were more isolated land masses. During periods when islands were more isolated, avian colonisers faced new selective pressures, new habitats and niches, leading to the evolution of island endemics, a process that probably took longer in some groups than in others.

On Cuba, a unique endemic family (Teretistridae) and eight endemic genera occur: *Cyanolimnas* (rail), *Starnoenas* (quail-dove), *Margarobyas* (owl), *Xiphidiopicus* (woodpecker), *Ferminia* (wren), *Teretistris* (Cuban warblers), *Torreornis* (sparrow), and *Ptiloxena* (blackbird), with a total of 29 endemic species suggesting a long isolation from continental lineages.

VEGETATION AND HABITATS

Knowledge of Cuba's habitat types can assist greatly in finding its birds. The diversity of Cuban vegetation is indicated by the existence of about 30 types of habitat, making the Cuban archipelago a conservation hotspot. There is a total of 7,020 species of vascular plants, of which some 6,000 are flowering plants and about 50 per cent are endemic to the island. Among the endemic species, 15 per cent are found mainly at low elevations, with approximately 75 per cent found in highland regions. The Nipe–Sagua–Baracoa mountain system has the highest percentage of endemics. Generally speaking, semi-deciduous forests account for about 43 per cent of all forested areas, mangroves 31 per cent, and pine forests, relicts from past ice ages, 12 per cent. The vegetative diversity of the Cuban archipelago has been impoverished: 500 years ago, forest accounted for 95 per cent, but today the figure is just 14–16 per cent.

Extensive rainforests cover the eastern mountains, whereas the south-eastern coastal ranges from Guantánamo to Maisí are extremely arid. Savannas altered by agriculture are often extensive, and biologically quite distinct. Noteworthy are the sandy savannas of the Isle of Pines and Pinar del Río and those on the large stretches of flat terrain scattered between the Havana region and the easternmost mountain ranges with their many Royal Palms, *Roystonea regia*. Swampy vegetation is abundant on the Zapata Peninsula.

The following is a summary of some important bird habitats shown on the map (the vernacular name is shown in parenthesis), we highlight major examples in the Cuban archipelago, including several important habitat sub-types.

Topographic map of Cuba.

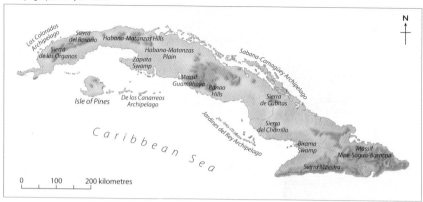

Vegetation on the Cuban archipelago.

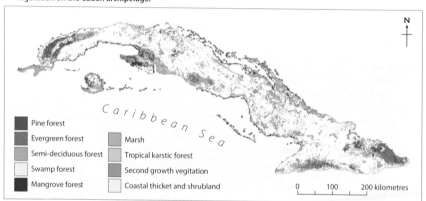

Cloud forest, Pico Turquino.

TERRESTRIAL HABITATS

Cloud forest typically occurs from 1,600–1,900m above sea level, although this formation is sometimes found at lower elevations. Annual precipitation range is 170–300cm. Cloud forest has no deciduous elements and the tree height averages 12m. Vegetation is almost continuously shrouded in mist from the near-constant cloud cover, and the high humidity supports luxuriant growths of arboreal and terrestrial ferns, mosses, liverworts and orchids, as well as a grass strata. Examples of cloud forest occur in the highest mountains of Sierra Maestra, Sierra de Imías, Sierra del Purial and other eastern mountain ranges, most notably at Pico Turquino. Indicator plant species found in cloud forest are *Magnolia* spp., *Cyrilla silvae*, *Graffenrieda rufescens*, *Ilex macfadyenii*, *Brunellia comocladifolia*, *Tabebuia shaferi*, *Garrya fadyenii*, *Torralbasia cuneifolia* and *Weinmannia pinnata*.

Rainforest is characterised by emergent deciduous trees, an annual precipitation of 120–300cm, and no dry season. Epiphytes and arboreal ferns are abundant. Three sub-types of rainforest are recognised on Cuba:

Rainforest, Humboldt National Park, Guantánamo.

a) **Lowland rainforest** extends from the coast to as high as 400m above sea level, with three arboreal strata, the tallest of which attains 35m. Annual rainfall is 160–300cm. In some localities of Moaëse phytogeographic district, rainfall may reach 500cm. Examples of lowland rainforest are found at Moa and Toa (e.g. the mouths of the rivers Jaguaní, Moa and Toa; Mal Nombre, a locality in Guantánamo province). In Guantánamo province, lowland rainforest is present at Meseta Mina Iberia and El Yunque near Baracoa. Indicator species are: *Carapa guianensis*, *Calophyllum utile*, *Oxandra laurifolia*, *Bactris cubensis*, *Tabebuia dubia*, *T. hypoleuca* and *Protium fragrans*.

b) **Submontane seasonal tropical rainforest** is found at elevations of 200–800m in mountains throughout Cuba, with rainfall of 80–300cm. It is similar to semi-deciduous forest, but with a greater proportion of evergreen species. Examples include: in Pinar del Río province, on the north-eastern slope of Pan de Guajaibón, some areas of Sierra del Rangel and Sierra del Rosario; in Cienfuegos province, near Pico San Juan; and in Sancti-Spíritus province, in some areas of Banao and Pico Potrerillo. Indicator species include *Andira inermis*, *Bursera simaruba*, *Sideroxylon foetidissimum*, *S. salicifolium*, *Cupania* spp., *Clusia rosea*, *Roystonea regia* and *Chionanthus ligustrinus*.

c) **Montane tropical rainforest** occurs at elevations of 800–1,600m in areas with rainfall of 160–230cm. It is characterised by two strata, the tallest of which attains 25m. Examples can be observed in Cuchillas del Toa and Moa, as well as at several localities in Sierra Maestra, including Gran Piedra and Pico Mogote, Meseta de Mina Iberia, part of Meseta del Toldo and some wooded areas around La Melba. Typically, trees are covered with epiphytes (orchids and bromeliads) and arboreal ferns (e.g. *Byrsonima crassifolia* and *B. bucherae*). Indicator species include *Beilschmiedia pendula*, *Clethra cubensis*, *Guatteria blainii*, *G. moralesi*, *Matayba domingensis*, *Micropholis polita* and *Bonnetia cubensis*.

Semi-deciduous woodland originally covered the majority of the island, in flat and rolling regions, the lower slopes of mountains and other seasonally humid areas. Annual rainfall is 100–120cm. It is a broadleaf woodland habitat, with 40–65 per cent of the upper stratum comprising of deciduous trees, and an understorey that includes bushes, grasses, some epiphytes and abundant lianas (woody vines). Two sub-types of semi-deciduous woodland are recognised on Cuba. Below are the characteristics of both sub-types:

a) **Semi-deciduous mesophyllous woodland** occurs in savanna, with two strata of trees no more than 15–25m in height, with individual trees and palms emergent above the highest stratum. Tree leaves are typically longer than 13cm. This is the most extensive woodland on Cuba, but such habitats are mostly relicts of formerly more extensive areas and now mostly consist of fragmented and degraded areas. Annual rainfall is 120–160cm. Examples include: in Pinar del Río province, Guanahacabibes Peninsula, the lower parts of the limestone karst hills of Viñales, Mil Cumbres; Isle of Pines, the southern peninsula; most of the karstic region of Península de Zapata; typical of the lower parts of the Guamuhaya mountain range in three of the central provinces; in Ciego de Ávila province, parts of Sierra de Cunagua; and in Camagüey province, small areas of Sierra Cubitas and Najasa. Indicator species are *Cordia gerascanthus*, *Zanthoxylum martinicense*, *Z. fagara*, *Cordia collococca*, *Nectandra coriacea*, *Cedrela odorata*, *Ceiba pentandra*, *Trichilia hirta* and *Oxandra lanceolata*.

Burrowing Owl habitat on sandy savanna, San Ubaldo, Pinar del Río province.

b) **Semi-deciduous microphyllous woodland** occurs in savannas and hills, and is characterised by two strata of trees of no more than 4–10m and 12–15m tall, respectively, and by having leaves less than 6cm in length. Annual rainfall is 80–120cm. Examples include: Isle of Pines, the southern region; in Matanzas province, La Barranca forest, Martí; Villa Clara province, small areas of Monte Ramonal in Santo Domingo; and part of Birama-Monte Cabaniguán; in Granma province, north of the mouth of The Río Cauto; in Santiago de Cuba province, areas of Siboney-Justicí, Daiquirí-Verraco to near Laguna de Baconao; and in Guantánamo province, small areas of scrubby coastal and subcoastal vegetation from Tortuguilla to near Maisí terrace. Indicator trees are generally thorny and typically include *Brya ebenus*, *Bucida* spp., *Erythroxylum rotundifolium*, *Copernicia* spp., *Cameraria* spp., *Hypelate trifoliata*, *Guettarda elliptica*, *Peltophorum adnatum*, *Coccothrinax* spp. and *Guaiacum officinale*.

Evergreen forest has broadleaf tree species making up less than 30 per cent of vegetation. It is characterised by an understorey of brush and grass strata, almost no epiphytes and an increasing number of lianas, depending on the conservation state. Two sub-types are recognised on Cuba:

a) **Evergreen mesophyllous forest** (local name, *manacal*, mainly in eastern Cuba) is typically found at 300–800m; rainfall is 120–230cm. The majority of trees are evergreen, with deciduous species comprising less than 30 per cent in the highest strata (usually no more than 25m high, but with some emergents reaching 30m) and with leaves characteristically more than 13cm in length. Examples include: in Pinar del Río province, lower slopes of Pan de Guajaibón, Sierra Chiquita and other localities between Mil Cumbres and Sabanilla; in Matanzas province, some areas within Sierra de Bibanasí and south of Santo Tomás, Caleta Buena in Zapata Swamp; in the central provinces, some zones within the Guamuhaya mountain range; in Ciego de Ávila province, parts of Sierra de Cunagua; in Camagüey province, some zones within Sierra de Cubitas and Najasa; and in Santiago de Cuba province, several lowland and mid-elevation localities in Sierra Maestra, including La Bayamesa National Park to the zones of La Gran Piedra–Pico Mogote, among other areas. Indicator species are *Alchornea latifolia*, *Sapium laurifolium*, *Dendropanax arboreus*, *Calophyllum antillanum*, *Trophis racemose* and *Pseudolmedia spuria*.

b) **Evergreen microphyllous forest** (dry forest) generally occurs in lowlands and hills near dry coasts. Annual rainfall is 70–120cm. It is characterised by evergreen and deciduous trees with leaves 1–6cm in length, an upper stratum reaching 12–15m and a lower stratum of 5–10m. Vegetation characteristically includes thorny bushes, dry epiphytes and columnar cacti. Examples include: in Pinar del Río province, some areas on Guanahacabibes Peninsula; in Matanzas province, relict patches from Punta Guanos to Punta Rubalcaba, and near Playa Girón to Punta Mangles; in Cienfuegos province, a small area between Punta Gavilanes and Trinidad; in Villa Clara province, part of Cayo Santa María; in Ciego de Ávila–Camagüey provinces, restricted areas in Cayo Coco; and in Granma-Santiago de Cuba–Guantánamo provinces, throughout the dry ecosystems of the southern coast of the Oriente region, including from Cabo Cruz to Maisí. Indicator plant species are *Colubrina elliptica*, *Bourreria suculenta*, *Exostema caribaeum*, *Coccothrinax* spp., *Ocotea nemodaphne*, *Cynophalla flexuosa*, *Cordia sebestena* and *Coccoloba diversifolia*.

Pine forest (*pinar*) is needle-leaved evergreen woodland, with the tree strata dominated by species of the genus *Pinus*, reaching 25m. The shrub and herbaceous strata are well developed, with some epiphytes and climbing plants. Pine forests occur in eastern and western Cuba, as well as the Isle of Pines. In the west, the annual rainfall is 100–160cm. Examples include: in Pinar del Río province, at lower levels of the base of Sierra de la Güira, Cajálbana, Mil Cumbres, Matahambre; and Isle of Pines, Los Indios, Santa Bárbara. In the east, the annual rainfall is 120–300cm. Examples include: in Santiago de Cuba province, La Gran Piedra, La Bayamesa; in Holguín province, Mayarí, Sierra de Nipe, Moa, Meseta del Toldo; and in Guantánamo province, Ojito de Agua, Cupeyal del Norte, Nibujón and Baracoa. Among indicator species in western Cuba and the Isle of Pines are *Pinus caribaea*, *Pinus tropicalis*, *Matayba oppositifolia*, *Abarema obovalis*, *Amaioua corymbosa*, *Eugenia rigidifolia*, *Colpothrinax wrightii*, *Coccoloba* spp. and *Curatella americana*. Characteristic species in eastern Cuba are *Pinus cubensis*, *Jacaranda arborea*, *Eugenia pinetorum*, *Baccharis scoparioides*, *Abarema nipensis*, *Anemia coriacea*, *Pteridium caudatum*, *Suberanthus canellifolius* and *Pinus maestrensis*.

Pine forest, Cupeyal del Norte.

Tropical karstic forest.

Viñales Valley.

Tropical karstic forest (*mogote* vegetation complex) occurs in karst mountains and is primarily mesophyllic semi-deciduous forest. The complex is characterised by bushy vegetation, with succulent plants, epiphytes and lianas, and an arboreal stratum (palms) reaching 10m high. Annual rainfall is 150–220cm (Sierra de Los Órganos–Viñalense district). This vegetation formation occurs in western Cuba but also in the central and eastern parts of the island. A typical locality is the Viñales Valley of Pinar del Río province. Indicator species in western Cuba are *Agave* spp., *Actinostemon brachypodus*, *Pachira cubensis*, *Celtis iguanaea*, *Cuervea integrifolia*, *Erythrina cubensis*, *Gaussia princeps*, *Lantana strigosa*, *Plumeria sericifolia* and *Leucothrinax morrisii*. The common species in central and eastern Cuba are *Coccothrinax* spp., *Picrasma tetramera*, *Grisebachianthus carsticola*, *Garrya fadyenii*, *Gesneria cubensis*, *Hemithrinax compacta* and *Neobracea howardii*.

Montane serpentine shrubwood (*charrascal*) occurs in savannas, hills and mountains in eastern Cuba. Annual rainfall is 120–300cm depending on the phytogeographic district. The scrubby vegetation can reach 6m high, with emergent palms attaining 10m. Vegetative elements are mesophyllous and sclerophyllous, with thorny elements less abundant than in lowland serpentine shrubwoods, and with some herbs, climbing plants and dispersed epiphytes. The richest diversity of endemic species (35–40 per cent) is found this habitat. Examples include: Moa–Baracoa; in Sierra de Nipe, La Bandera, La Caridad and La Cueva; and Sierra Cristal. Among indicator plants are *Dracaena cubensis*, *Adenoa cubensis*, *Annona sclerophylla*, *Erythroxylum pedicellare*, *Mazaea shaferi*, *Byrsonima biflora* and *Coccoloba* spp.

Lowland serpentine shrubwood.

Lowland serpentine shrubwood (*cuabal*) occurs in savanna areas with low hills. It features dense scrubby vegetation reaching no more than 4m, with emergents no taller than 6m. Vegetation is thorny, sclerophyllous, nanophyllous and herbaceous, with dispersed epiphytes on ultramafic soils derived from serpentine metamorphic rocks. Annual rainfall is 80–160cm. Examples include: in Pinar del Río province, Cajálbana, in Havana province, Barreras, La Coca; in Matanzas province, Tres Ceibas de Clavellinas, Loma el Jacán and San Miguel de los Baños; in Villa Clara province, Motembo, around Los Caneyes, Cubanacán and Agabama; in Sancti-Spíritus and Ciego de Ávila provinces, Jatibonico–San Felipe; in Camagüey province, Sabanas de Cromo; and Holguín province. Among the indicator species are *Phyllanthus orbicularis*, *Neobracea valenzuelana*, *Mesechites roseus*, *Bursera angustata*, *Coccothrinax* spp. and *Erythroxylum alaternifolium*.

Coastal thicket comprises scrubby vegetation in coastal areas, and is characterised by stunted shrubs and emergent trees, with palms reaching up to 6m. Vegetation has deciduous elements, including sclerophyllous, microphyllous and nanophyllous, with leaf types less than 3mm in length. Succulents and thorny elements are abundant, and some palms, grass and vines occur. Annual rainfall is 100–140cm. Examples include: in Pinar del Río province, Guanahacabibes; in Matanzas province, Zapata Peninsula (near Playa Girón, near Guasasa); Cayo Santa María, on some rocky coasts; in Ciego de Ávila and Camagüey provinces, areas on the northern cays in the Sabana–Camagüey archipelago; in Holguín province, some coastal points of Guardalavaca and Caletones in Gibara; and in Guantánamo province, southern coast from Tortuguilla to Maisí, Baracoa (average rainfall 38–60cm) and Bahía de Taco. Indicator species include *Erythroxylum rotundifolium*, *Quadrella cynophallophora*, *Capparis grisebachii*, *Diospyros* spp., *Coccothrinax* spp., *Caesalpinia* spp., *Guettarda* spp. and *Maytenus buxifolia*. In some regions, particularly the southern coast of the eastern provinces, there are many species of succulents and spiny brush, notably *magueyes* and cacti, including *Harrisia eriophora*, *Dendrocereus nudiflorus*, *Pilosocereus* spp., *Consolea macracantha*, *Opuntia dillenii* and *Melocactus* spp.

Coastal thicket, Baitiquirí.

Riparian forest occurs along river and stream edges, in all provinces where natural vegetation and/or second growth is found along rivers, streams, springs and canals. It has a tree stratum of 15–20m, is composed of more sun-loving plant species, including palms and an understorey of shrubs, grasses, climbing plants and epiphytes. Annual rainfall is 70–120cm. Indicator species are *Calophyllum brasiliense*, *Lonchocarpus seriseus*, *Tabebuia angustata*, *T. leptoneura* and *Calophyllum rivulare*.

Lowland riparian forest, Zapata Peninsula.

Montane riparian forest, La Melba, Humboldt National Park.

WETLAND HABITATS

These are well represented on Cuba. There are the several natural wetlands, which include sandy coast and coral reefs ecosystems comprising 5,746km^2 and also inland ecosystems.

Swamp forest is characterised by broadleaf woodland, with deciduous elements, occupying zones flooded periodically or permanently in coastal regions. Annual rainfall is 120–170cm. The tree stratum is 8–15m, with a maximum of 20m. This habitat includes mangrove, epiphyte and fern elements, with

Thatch palms in sandy coast vegetation, Cayo Paredón Grande, Ciego de Ávila province.

Shorebird habitat on tidal sand bank, Cayo Coco, Ciego de Ávila province.

grasses and herbaceous growth on soils rich in organic material. Example localities include: in Pinar del Río province, northern Guanahacabibes Peninsula; southern Mayabeque near Surgidero de Batabanó; in Matanzas province, Zapata Peninsula, Ciénaga de Majaguillar; southern Isle of Pines; in Villa Clara province, Ciénaga de Jumagua (Sagua la Grande); in Ciego de Ávila province, wetlands of northern Ciego de Ávila–Ciénaga de Cunagua; and in Granma province, Ciénaga de Birama–Monte Cabaniguán. Indicator species are *Bucida palustris*, *B. buceras*, *Annona glabra*, *Tabebuia angustata*, *Hibiscus elatus*, *Morella cerifera*, *Sabal maritima* and *Copernicia* spp.

Rocky coast vegetation complex includes communities in open areas on high rocky coasts, with small to large succulents, small bushes up to 2m high and grasses. Annual rainfall is 100–140cm. Example localities include: in Pinar del Río province, Guanahacabibes Peninsula; southern part of Isle of Pines; north-east of Havana; in Matanzas province, Zapata Peninsula, Archipelago de los Canarreos; in Camagüey and Ciego de Ávila provinces, northern and southern cays; and in Granma, Santiago de Cuba and Guantánamo provinces, Cabo Cruz to Maisí. This complex is characterised by *Borrichia arborescens*, *B. cubana*, *Euphorbia mesembryanthemifolia*, *Erithalis fruticosa*, *Flaveria linearis* and *Strumpfia maritima*.

Sandy coast vegetation complex comprises open communities on sandy areas. Annual rainfall is 100–140cm. It includes grassy plants, palms and dispersed trees; mangroves or sea grape (*Coccoloba uvifera*) are the usual arboreal species. Example localities include: in Pinar del Río province, Guanahacabibes Peninsula; southern part of Isle of Pines; north-east of Havana; in Matanzas province, Zapata Peninsula, Archipelago de los Canarreos; and in Camagüey and Ciego de Ávila provinces, northern and southern cays. Among the indicator species are *Canavalia rosea*, *Cenchrus tribuliodes*, *Diodia serrulata*, *Erithalis fruticosa*, *Ernodea littoralis*, *Ipomea pes-caprae*, *Scaevola plumieri*, *Stemodia maritima* and *Uniola virgata*.

Coastal areas include fringing reefs and the coastline of sandy beaches with no terrestrial vegetation.

Sandy coast vegetation complex, Maisí, eastern Cuba.

SALT AND BRACKISH WATER COMMUNITIES

Mangrove forest is broadleaf evergreen woodland alongside low muddy coastal or estuarine areas with a high level of salinity. Mangrove forest covers 4.8 per cent of the entire area of the Cuban archipelago and represents 26 per cent of forested land. Annual rainfall is 120–170cm in the Zapatensis phytogeographic district. Trees can reach 15–25m, are adapted to submersion and have specialised roots (prop roots and pneumatophores). No bush stratum is present, but grasses with succulent leaves, ferns and climbing plants (creepers) are well represented. Mangrove forest is best developed in the following areas: the southern coast of the Isle of Pines and its cays; in Pinar del Río province, along the northern coast from Las Pozas to Pretiles, and small areas in northern Guanahacabibes; in Mayabeque, all along the southern coast from Ensenada de la Broa; in Matanzas province, on the Zapata Peninsula and Hicacos Peninsula; in Villa Clara province, at Carahatas in Sagua la Grande, northern cays; in Ciego de Ávila province, some parts of the northern and southern cays; in Camagüey province, some parts of the northern and southern cays; and in Granma province, the delta of the Cauto River and Monte Cabaniguán. Indicator species are *Rhizophora mangle*, *Avicennia germinans*, *Laguncularia racemosa*, *Conocarpus erectus*, *Acrostichum aureum*, *Dalbergia ecastaphyllum*, *Batis maritima* and *Rhabdadenia biflora*.

Mangrove forest, La Salina, Zapata Peninsula.

Salt ponds are coastal wetlands periodically infused with saltwater during very high tides or storms. Most vegetation consists of herbaceous species and succulent plants that are highly tolerant of salt. Indicator species include *Batis maritima*, *Chloris sagrana*, *Distichlis spicata*, *Fimbristylis spathacea* and *Sarcocornia perennis*.

FRESHWATER COMMUNITIES

Marsh (*herbazal de ciénaga*) consists of seasonally flooded grassland, and typically remains waterlogged at all times. Annual rainfall is 120–170cm in the Zapatensis phytogeographic district. The grassland can reach 1.5–2.0m high. This habitat is typically found on Zapata Peninsula, in Birama Swamp and Lanier Swamp. Indicator species include *Cyperus* spp., *Echinodorus* spp., *Eleocharis cellulosa*, *E. interstincta*, *Panicum dichotomiflorum* and *P. lacustre*. In areas that are permanently flooded and have an accumulation of peat in the soil, the following species may be present: *Centella erecta*, *Cladium jamaicense*, *Thelypteris confluens*, *Typha domingensis*, *Erianthus giganteus* and *Cyperus giganteus*.

Zapata marsh (eastern).

Zapata marsh (western); Zapata Wren habitat.

Freshwater wetland, San Ubaldo, Pinar del Río province.

Marshlands besides rivers are lowland habitats, including those along the largest rivers, the Cauto and Toa, in eastern Cuba. Among the plant species present are *Arundo donax*, *Bambusa vulgaris*, *Cyperis heterophyllus*, *C. surinamemnsis* and *Gynerium sagittatum*.

Freshwater lagoons are characterised by floating and rooted species of plants. Among the floating species are *Azolla caroliniana*, *Eichhornia* spp., *Lemma minuta*, *Pistia stratiotes*, *Salvinia auriculata* and *Utricularia* spp. Among the rooted species are *Brasenia schreberi*, *Cabomba furcata*, *Hydrocotyle umbellata*, *Nymphaea* spp. and *Potamogeton* spp.

Anthropic wetlands are freshwater habitats, including rice fields and reservoirs, that are artificially created and maintained. The plant communities include many species that occur in natural freshwater aquatic habitats.

Second-growth vegetation occurs where the original natural vegetation has been altered by human activity or natural events (e.g. hurricanes). The original vegetation has been replaced by other species and associations, with the structural complexity related to the levels of successional development in three categories of associations:
a) **Second-growth forest** has arboreal, brush and grass strata, with an abundance of creeping vines and widely dispersed trees.
b) **Bushland** is characterised by an abundance of shrubs, grass and creeping vines, with widely dispersed trees.
c) **Grasslands** have a well-developed herbaceous stratum; there are dispersed trees, palms, bushes, vines and climbing plants.

Sandhill Crane habitat in second-growth vegetation, Zapata Peninsula.

Agricultural lands include all those areas cleared for agriculture, whether for large-scale farming or for subsistence agriculture, at high or low elevation.

Urban areas are extensively human-modified habitats, where primary habitats are mainly replaced with exotic plants in parks and gardens, artificial constructions and any other type of human activity.

HISTORY OF ORNITHOLOGY IN CUBA

The earliest descriptions of endemic species, Blue-headed Quail-Dove and Cuban Bullfinch, were by Linnaeus in 1752, and of Cuban Grassquit by Johann Friedrich Gmelin in 1789. Serious ornithological studies began in the first half of the 19th century. In 1827, Temminck described and illustrated some species from Cuba in a catalogue in 1825. Pablo Guillermo de Würtenberg collected birds and published *Viaje a América septentrional* in 1835. In 1839, an ornithological section was included in a compendium, *Historia física, política y natural de la Isla de Cuba*, by Alcide d'Orbigny, published by Don Ramón de la Sagra. In 1847, he also published a dissertation, *Ornitología de la India Occidental*. In 1848, Andrés Poey y Aloy, published a catalogue of Cuban birds numbering 208 species *in Memorias de la Real Sociedad Económica de la Habana*. In 1850, Juan Lembeye produced *Aves de la Isla de Cuba*.

Johannes Gundlach arrived in Cuba in 1838, becoming one of its most distinguished naturalists and ornithologists. He discovered several Cuban endemics, including Gundlach's Hawk, Gray-fronted Quail-Dove, Bee Hummingbird, Cuban Vireo, Cuban Gnatcatcher and Oriente Warbler, and also discovered other Antillean endemics such as Antillean Nighthawk, La Sagra's Flycatcher, Bahama Mockingbird and Olive-capped Warbler; he described some of these species. Gundlach wrote two major treatises on Cuban birds, *Contribución a la Ornitología Cubana* (1876, 1893). The majority of endemics were described between 1811–60, though Giant Kingbird (a *de facto* endemic) was not described until 1887. The last endemic species to be found were Zapata Wren, Zapata Sparrow and Zapata Rail in 1926–27 by the Spanish entomologist, Fermín Cervera; the first of this trio was described by Thomas Barbour and the other two by Barbour and Peters.

In the first half of the 20th century, important ornithological contributions were made by W. E. Clyde Todd, who wrote *The Birds of the Isle of Pines* (1916); Thomas Barbour, a zoologist from the Museum of Comparative Zoology, Harvard University, who wrote *Birds of Cuba* (1923) and *A naturalist in Cuba* (1946); and James Bond, Curator at the Academy of Natural Science, Philadelphia, who published *Birds of the West Indies* (1936, with several later editions and supplements to 1987).

Many people have contributed to Cuban ornithology, including the North American entomologist, Stephen Bruner, and Charles Ramsden in Santiago de Cuba studied birds in the eastern part of the country, collecting several species that are almost extinct today. Amateurs played major roles, including the collectors Gastón Villalba, Cleto Sánchez and José Hernández Bauzá, an excellent taxidermist who prepared specimens of mounted birds, and eggs, for two important collections. Pastor Alayo Dalmau and Abelardo Moreno Bonilla also made significant contributions. Important researchers have included Robert Ridgway, Wells W. Cooke, Outram Bangs, Walter R. Zappey, Pelegrin Franganillo, Storrs Olson, Kenneth C. Parkes, George Watson, Alexander Wetmore, Peters Friedman, William Suárez, George Wallace, José Morales, Pedro Blanco, Alejandro Llanes, Martín Acosta, Lourdes Mujica, Orlando Torres, the late Denis Dennis, Ariam Jiménez, Hiram González, Barbara Sánchez, Daisy Rodríguez and Freddy Santana, among others.

Cuban zoologists Orlando H. Garrido and Florentino Garcia Montaña produced *Catálogo de las Aves de Cuba* (1975) for the Felipe Poey museum. Garrido and Arturo Kirkconnell wrote *A Field Guide to the Birds of Cuba* (2000, 2011). Other important contributions have included *Aves Acuáticas en los humedales de Cuba* (Mujica *et al.* 2006); *Important Bird Areas (IBAs) in Cuba* (2010); and the multi-author *Libro rojo de los Vertebrados de Cuba* (2012). However, in the last three decades the most important and long-term contribution to the natural history of Cuban endemic birds was by James W. Wiley (1943–2018).

Bee Hummingbird.

THE AVIFAUNA

BREEDING BIRDS

A total of 156 species are known to have bred regularly on Cuba. The breeding season is generally March–July, but a number of species breed outside this period and some appear to nest year-round. Changes in the breeding season of several species have been noted, beginning earlier in late January or prolonged until August. The breeding season appears to be related to the onset of the wet season (rain cycles can also change in relation to the altitude and region) and is interconnected with food availability (abundance of insects, blooming and fruiting of certain trees), required to provide the energy levels for adults and chicks during the breeding period.

LANDBIRDS

A total of 96 species (61 per cent) of native (indigenous) species are landbirds and 44 of these (46 per cent) breed in forest. Among the 29 endemic species, a total of 24 breed in forest regularly; the exceptions are Zapata Rail, Zapata Wren, Cuban Gnatcatcher and Red-shouldered Blackbird. In the case of Cuban Grassquit, Zapata Sparrow and Cuban Vireo, the breeding range can expand into bushes or scrub.

Nests constructed by forest birds vary from simple to more elaborate and complex structures, and are usually attached to or suspended from trees. Many passerines (such as warblers and vireos) build cup-shaped nests, others (e.g. quail-doves, pigeons, kingbirds and mockingbirds) build very simple stick platforms, and others (e.g. Cuban Oriole, Zapata Wren, Cuban Bullfinch and Cuban Grassquit) construct complex globular nests.

Primary cavity nesters, including all woodpeckers, expend substantial energy excavating their nest hole. Despite this initial investment, these sites can be reused by woodpeckers over several years, requiring only cleaning and enlargement of the interior dimensions. Importantly, American Kestrel, Cuban Trogon, Cuban Parrot, Cuban Parakeet, Cuban Martin, Cuban Pygmy-Owl and Bare-legged Owl reuse the cavities built by woodpeckers. In addition to woodpeckers, other species, such as Cuban Tody, excavate their own cavities in soft substrates such as mud, sand, rotten wood in dead or living trees, or use existing natural holes. Other species add material to natural cavities, e.g. Red-legged Thrush and Cuban Blackbird (the latter also uses woodpecker cavities). Endemic Cuban Solitaire, Cuban Green Woodpecker, Cuban Tody and Cuban Parakeet may also breed in natural cavities in limestone cliffs, as do Barn Owl, White-collared Swift and Black Swift.

Red-legged Thrush, western subspecies.

Only one forest species, the cryptic Cuban Nightjar, breeds on the ground in litter on the forest floor. Antillean Nighthawk, another ground-nesting member of this family, nests in open terrain with pale (often stony or sandy) ground. Three endemic species are confined to marsh vegetation: Zapata Wren (globular nest), Zapata Rail (unknown) and Red-shouldered Blackbird (cup-shaped nest). Nesting materials also vary, with a broad spectrum of items used, such as sticks, grasses, leaves and hair. Species such as Cave Swallow construct their nests with mud and saliva. The Bee Hummingbird's nest is the smallest and possibly the most complex among Cuban birds. This species uses grass, lichen and spider webs; only the female builds and tends the nest, as is typical of the hummingbird family.

A total of 16 species visit Cuba to breed, mostly from southern regions: Yellow-billed Cuckoo, Antillean Nighthawk, Wilson's Plover, Snowy Plover, Sooty Tern, Bridled Tern, Least Tern, Roseate Tern, Common Tern, Sandwich Tern, White-tailed Tropicbird, Audubon's Shearwater, Gray Kingbird, Black-whiskered Vireo, Cuban Martin and Cave Swallow.

WATERBIRDS

A total of 59 (39 per cent) species breed in wetlands, of which 21 breed along Cuban coasts. The only endemic waterbird is the Zapata Rail. All the breeding plovers (Killdeer, Wilson's Plover and Snowy Plover) nest on the ground in open habitats. Many seabirds select their nest-site based on factors such as substrate and vegetation cover, temperature, distance from feeding areas and presence of predators.

Only ten seabirds breed in the Cuban archipelago, nesting on the rocky or sandy substrate of cays, cliffs and beaches, mainly from May–July: Laughing Gull, Brown Noddy, Sooty Tern, Bridled Tern, Least Tern, Gull-billed Tern, Common Tern, Roseate Tern, Royal Tern and Sandwich Tern. Six species of ducks breed on Cuba, from April/May through the summer and possibly year-round: West Indian Whistling-Duck, Fulvous Whistling-Duck, Wood Duck, White-cheeked Pintail, Masked Duck and Ruddy Duck.

Mangroves are important coastal wetland ecosystems for bird reproduction because of their abundant food sources, secure nesting sites and relative protection against land predators. Among the most important Cuban mangrove species is the American Flamingo, with around 130,000 individuals. A colonial breeder, its largest colony is at the mouth of The Río Máximo in Camagüey province, with c. 45,000 pairs. Clapper Rail also breeds in this habitat, as do Pied-billed Grebe and Least Grebe, which breed almost year-round (and also in freshwater wetlands).

Both members of the family Recurvirostridae breed: Black-necked Stilt and American Avocet, though the latter has only been found to nest very recently. Of the 26 species of wader on Cuba, only Snowy Plover, Wilson's Plover, Killdeer and Willet breed. Nesting only on offshore cays, mostly from

Least Tern, non-breeding.

Matanzas to Camagüey province, are Magnificent Frigatebird almost year-round (few breeding data are available from Cuba), and two species of cormorants, Double-crested and Neotropic. Brown Booby also breeds in small numbers on some offshore cays, as do Anhinga and Brown Pelican.

As might be expected given the abundance of wetlands on Cuba, there are 11 breeding members of the heron family. Egret breeding seasons are generally the longest of any in this family, ranging from 6–11 months. Little Blue Heron and Reddish Egret have very long breeding seasons, while Least Bittern, Great Blue Heron and Cattle Egret are among those with the shortest. The most important heron breeding colonies are in Matanzas province east to Birama in Granma province, including the cays.

Seven species of rail breed in freshwater wetlands on Cuba. The Zapata marshland is the only site for the Critically Endangered Zapata Rail; breeding data unknown. It also holds the most important breeding populations of Spotted Rail and Yellow-breasted Crake, King Rail, Purple Gallinule, Common Gallinule, American Coot and Limpkin. Northern Jacana is an uncommon resident breeder, its population decline throughout the island possibly being due to the introduction of exotic catfish. Two hawks breed exclusively in freshwater wetlands: Snail Kite and Osprey.

THREATENED AND RESTRICTED ENDEMIC SPECIES AND SUBSPECIES

The majority of Cuban endemics have a fragmented breeding distribution throughout the main island; fewer breed on the Isle of Pines (Isle of Youth) and the northern and southern cays. The most restricted species are Zapata Rail and Zapata Wren, both relict populations restricted to western Zapata Peninsula. Zapata Rail and Cuban Kite are the most threatened species on Cuba. Cuban Kite is found in the Nipe–Sagua–Baracoa mountain range. Another locally restricted species is Cuban Palm Crow, currently found near Trinidad, central and southern Camagüey. Zapata Sparrow occurs as three subspecies restricted to Zapata Peninsula, Cayo Coco, Cayo Romano and south-east Guantánamo. Other species, such as Fernandina's Flicker, are widespread but locally distributed. The flicker is a locally threatened species with the main populations from eastern Pinar del Río province to Las Tunas and Birama Swamp, although most individuals occur in Zapata Peninsula. Another restricted species is Red-shouldered Blackbird, found in the central and western section of the main island and the Isle of Pines. Some endemic subspecies are confined to the Isle of Pines: Cuban Green Woodpecker, West Indian Woodpecker and Cuban Trogon.

Snail Kites in flight.

Giant Kingbird is widespread throughout the island although very local, with most individuals located in southern Camagüey province. Cuban Gnatcatcher is mostly restricted to coastal areas in the central provinces and eastern Cuba, rarely inland. Cuban Solitaire breeds in the mountains of western and eastern Cuba; it is absent from the central mountain ranges. Cuban Parakeet is found mostly from Zapata Peninsula to the eastern mountain ranges of Cuba. The most important breeding populations of Blue-headed Quail-Dove are on Guanahacabibes Peninsula, Sierra de los Órganos, Sierra del Rosario and Zapata Peninsula.

The breeding areas of Sandhill Crane are very localised: Isle of Pines, Zapata Peninsula and scattered locations in Camagüey province. Sharp-shinned Hawk is very local, with the majority breeding in the mountains of the eastern and western provinces. If it is still extant, Ivory-billed Woodpecker is restricted to Humboldt National Park, Guantánamo province, but it may be extinct. Thick-billed Vireo is located in some northern cays; this species is highly threatened due to the impact of Hurricane Irma in September 2017 and new development. Red-legged Thrush has two very distinct subspecies: one widespread population in the western and central part of Cuba and the cays, and the second restricted to eastern Cuba.

MOULTS

Feathers undergo changes throughout the year, from worn and faded to bright and new following a full moult. Many species assume these changes gradually. Others have completely different plumages in the non-breeding and breeding seasons and may also show intermediate stages while in transition between the two, e.g. Black-bellied Plover and Rose-breasted Grosbeak. A full or partial moult to replace worn feathers usually occurs twice a year including the post-breeding moult, which may take place before migration, at migration stops or on arrival in the wintering grounds. The moult from immature to adult can result in a brighter but similar plumage, e.g. Ovenbird, or one distinct from the immature phase, e.g. Little Blue Heron.

Most juveniles have similar but duller plumages than adults, with paler feather tips and fringes, as in doves and many shorebirds, while others have distinctive field marks, e.g. Western and Least Sandpipers have short-lived bright cinnamon feathers on the face and scapulars, an important identification tool in early autumn. In most passerines, juvenile plumage lasts until late summer; the first-winter plumage follows and remains until the first spring when it is replaced by the first-summer plumage, which is often similar to the adult breeding plumage. The first adult non-breeding plumage is assumed through the second autumn and winter and is then replaced by adult breeding plumage the following spring. In some species, replacement occurs within a shorter period. In some seabirds, e.g. gulls, there are juvenile, first-, second-, third- and sometimes fourth-year plumages before adulthood. Other species assume adult plumage gradually, e.g. Red-footed Booby over three years.

Red-breasted Grosbeak, breeding male.

Thus, between autumn and spring different individuals of a single species may exhibit juvenile, immature, male, female, breeding and non-breeding plumages as well as transitional plumages. It takes many years of expertise to distinguish between the dull autumn plumages of non-breeding and immature birds. In autumn and winter, the majority of migrants are in non-breeding and immature plumage, which is dramatically different from the summer breeding plumage, especially in some shorebirds and warblers.

MIGRATION

The geographical position of the Cuban archipelago between the Florida Peninsula and the other Antillean islands and Central and South America makes Cuba an important wintering ground and stopover site for hundreds of thousands of birds migrating from North America to latitudes south of their breeding grounds. Cuba's land mass (with about 50 per cent of the total land area in the Caribbean), topography, habitat diversity and the fact that it is part of a chain of islands (Greater and Lesser Antilles) have all contributed to its importance for migrating birds. Migration studies have identified several different migration routes in North America. Two have a direct impact on the Cuban archipelago: the Atlantic coast corridor by which most birds arrive from Florida, and the Mississippi corridor across the Gulf of Mexico to the Yucatán Peninsula and then onward through Central and South America. The Atlantic coast corridor has the greatest influence, taking birds over the Bahamas and Cuba.

Autumn migration Between mid-July and May bird diversity is greatly enriched by Nearctic migrants, with 138 species overwintering on Cuba for 7–10 months and 44 passage migrants remaining only for a brief stopover and wintering farther south, e.g. Wilson's Warbler and Prothonotary Warbler. Autumn migration extends from early July to late November, with several peaks in September, October and November depending on weather conditions en route and the timings when leaving the breeding grounds. Warblers, vireos, ducks, waders and herons migrate across the island, with a tendency for some to migrate to certain regions. The largest waves of mixed flocks of nocturnal migrants arrive along northern-western and central coasts of Cuba. Among the first species to arrive are Louisiana Waterthrush, Northern Waterthrush and Least Sandpiper; next are Prairie Warbler, Yellow-throated Warbler and Black-and-white Warbler. A steady flow of birds continues into November with a few later arrivals occurring into early December.

Spring migration Nearctic species from the south arrive along the entire southern coast, with the greatest numbers in the central and western part of the Cuban archipelago, i.e. from Las Tunas and Holguín provinces west to central Pinar de Río province. Migration begins as early as late January, intensifying in March and the first two weeks of April. In May and June it is rare to see wintering migratory birds and most are late passage migrants from further south. Most summer visitors are Neotropical species that arrive on Cuba from southern regions to breed; some are abundant and evenly distributed throughout the country (Antillean Nighthawk, Gray Kingbird, Black-whiskered Vireo), though Cuban Martin seems to be restricted by the availability of nesting sites. For some of the migratory species, e.g. warblers, Cuba was or is the only major wintering ground. Bachman's Warbler winters nowhere else. For others, such as Palm Warbler and Black-throated Blue Warbler, Cuba is simply the most important wintering site in the West Indies. Two migratory species, although widely distributed, show a strong preference for mountain habitats, e.g. Black-throated Blue Warbler in the highest

Black-whiskered Vireo.

mountains of the Sierra Maestra and surrounding mountain ranges (Nipe–Sagua–Baracoa). It occurs from sea level to at least 1,300m, where it is very common. Bicknell's Thrush only winters in the West Indies on Cuba, Hispaniola and Puerto Rico; on Cuba, it is a very rare winter visitor to the highest mountains in the eastern part of the island, Pico Turquino, Pico Suecia, Pico Cuba and La Bayamesa.

REGULAR MIGRATORY LANDBIRDS

The most important sites for overwintering landbirds are those places with the greatest diversity of native species and are usually in areas of conservation importance such as the Zapata Peninsula, Guanahacabibes Peninsula, Sierra de Guaniguanico, surrounding cays and the major mountain ranges of Nipe–Sagua–Baracoa.

Three species of falcon are regular migrants and overwinter: American Kestrel, Peregrine Falcon and Merlin. Sharp-shinned Hawk is a regular winter migrant and an uncommon migrant through North America to the West Indies. Other regular wintering migrant raptors include Broad-winged Hawk and Northern Harrier; the former migrates mostly through western Cuba, while the latter winters mostly in central and western Cuba. Swallow-tailed Kite usually arrives in the same areas of Cuba, before continuing on passage to the Yucatán Peninsula.

In the cuckoo family, Yellow-billed Cuckoo is both a summer breeding visitor and passage migrant. In the nightjar family, Common Nighthawk is a regular passage migrant. Antillean Nighthawk is a common summer breeder and passage migrant, and is thought to winter in South America. Among the swifts, only Chimney Swift is an uncommon passage migrant. There are eight migratory swallows. Tree Swallow is the most common winter visitor and passage migrant. Purple Martin migrate, through the Bahamas, Cuba and Cayman Islands; on Cuba it is a common passage migrant. Cave Swallow is a summer breeding visitor. Northern Rough-winged Swallow is mainly a passage migrant but possibly also a winter visitor. Barn Swallow is a common passage migrant. Cuban Martin is a common summer breeding visitor. Yellow-bellied Sapsucker, the only migratory woodpecker on Cuba, is a common winter visitor and passage migrant.

A total of 13 migratory tyrant flycatchers occur. One, Gray Kingbird, is a common summer breeding visitor, whose autumn migration is along the southern coast through the eastern part of the island. Eastern Wood Pewee is a regular visitor in winter. The migrant Gray Catbird is one of the most common winter visitors and passage migrants, and Cuba is the most important wintering ground for this species in the Greater Antilles. Nine migratory thrushes are recorded, eight being mostly rare passage migrants. Seven migratory vireos occur. Black-whiskered Vireo winters in Amazonia; on Cuba, it is a very common summer breeding visitor in the western mountains. White-eyed Vireo and Yellow-throated Vireo are common winter visitors and passage migrants, mainly in western and central Cuba. Red-eyed Vireo is a common passage migrant. Blue-gray Gnatcatcher is a common winter visitor.

The warbler family, from which 40 species have been recorded, is the most conspicuous group of migratory landbirds arriving on Cuba. Sixteen species are regular in winter. Among the most common are Palm Warbler, American Redstart, Black-and-white Warbler, Northern Parula, Black-throated Blue Warbler, Prairie Warbler, Common Yellowthroat, Cape May Warbler, Yellow-throated Warbler, Northern Waterthrush and Ovenbird. Some of the warblers arrive in mid-July, but peak arrivals are from mid-September to October. A total of 12 species of warbler are mainly passage migrants, although some individuals may winter: Prothonotary Warbler, Tennessee Warbler, Hooded Warbler, Yellow Warbler, Chestnut-sided Warbler and Blackpoll Warbler are regular passage migrants.

Chestnut-sided Warbler, breeding male.

Among the cardinals, only two species are regular visitors: Summer Tanager, a winter resident and passage migrant, and Scarlet Tanager, a passage migrant. A total of eight sparrows occur, of which five are regular winter visitors: Savannah Sparrow, Grasshopper Sparrow, Clay-coloured Sparrow, White-crowned Sparrow and Lincoln Sparrow. Most migrate through the western section of the island. Other regular visitors are Blue Grosbeak, Indigo Bunting, Rose-breasted Grosbeak and Painting Bunting. Among the blackbird family there are six migratory species, of which three are regular passage migrants with Baltimore Oriole and Bobolink the most common.

REGULAR MIGRATORY WATERBIRDS

Waterfowl and shorebirds migrate by both day and night, partly depending on weather conditions. Sandpipers, herons and ducks make up the largest percentage in aquatic ecosystems, visiting beaches, mangroves, marshes, marshy plains, lakes, dams and agro-ecosystems such as rice. On Cuba, most waders are observed on mudflats and shallow lagoons of brackish water in mangrove habitats or on remote sandy beaches and on rice fields. Most mudflats are located in the southern part of the Cuban archipelago: Zapata Swamp (Matanzas province), Lanier Swamp (Isle of Pines), Birama Swamp (Granma province) and Bahia de Guadiana (Guanahacabibes) in Pinar de Río province.

THE MOST COMMON VISITORS IN WINTER AND/OR ON PASSAGE

Among the duck species are Blue-winged Teal, Northern Shoveler, American Wigeon, Ring-necked Duck, Lesser Scaup, Red-breasted Merganser and Ruddy Duck. Among the rails are American Coot (which also has a resident breeding population) and Sora.

Shorebirds occur in suitable habitats throughout most of the island. The most common winter visitors are Black-necked Stilt (also a breeder), Black-bellied Plover, Wilson's Plover (a summer breeding resident and passage migrant), Semipalmated Plover, Killdeer, Ruddy Turnstone, Least Sandpiper, Semipalmated Sandpiper, Short-billed Dowitcher, Wilson's Snipe, Spotted Sandpiper, and both Greater and Lesser Yellowlegs (which are recorded year-round). Willet is a summer-breeding resident, winter visitor and passage migrant. Stilt Sandpiper is a passage migrant. Among the family Laridae, Ring-billed Gull and Caspian Tern are common winter visitors and passage migrants. Laughing Gull and Royal Tern occur in winter and on passage, and both species are found commonly year-round and breed on Cuba.

Killdeer.

VAGRANTS AND RARE VISITORS

A total of 67 species are currently considered to be vagrant, accidental or occasional visitors. The most recently recorded is White-faced Ibis (see Appendix A).

INTRODUCED AND COLONISING SPECIES

Six species are confirmed as introduced and another two were possibly introduced (see Appendix B).

CONSERVATION

CONSERVATION LAW ON CUBA

Environmental considerations were largely ignored on Cuba for almost 200 years. Only in the last few decades, with the enactment of Law 33 (the Law on Environmental Protection and the Rational Use of Natural Resources) on 10 January 1981, have environmental laws and regulations begun to play a very small role in guiding the development of natural resources exploitation and the ecology of the island. Law 33 was passed in order to 'establish the basic principles to conserve, protect, improve and transform the environment and the rational use of natural resources, in accordance with integral development policies' of the Cuban government, and 'with the objective of the best utilisation of the national productive potential'. Law 33 is divided into four main chapters: chapter one covers the main concepts of the Law; chapter two covers specific areas of the Law and the fundamentals for the use, protection and rehabilitation of water, soil, mineral resources, etc.; chapter three covers the organisation of the government entity responsible for the Law – the Comisión Nacional de Protección del Medio Ambiente y Conservación de los Recursos Naturales; and chapter four is an attempt to legislate a system of fines for violating the Law, including a mechanism to insure that it is obeyed.

PROTECTED AREAS

The Sistema Nacional de Áreas Protegidas de Cuba (SNAP) cover approximately 20.2 per cent of the country, including the marine insular platform to a depth of 200m, with 17.2 per cent of the land and 25.0 per cent of the marine platform administered by the Sistema Nacional.

Currently 211 protected areas have been recognised for their value for conservation purposes under one of the management categories established for Cuba; 77 areas have been identified as of national significance and 112 of local significance. Coordinated by SNAP as part of BirdLife International's Caribbean Programme, 28 Important Bird Areas (IBAs) were identified, mapped and documented for Cuba.

Map of protected areas on Cuba.

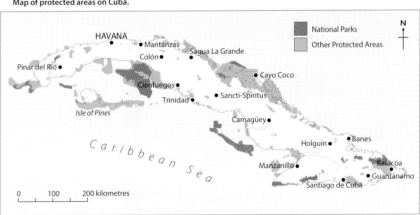

ADVERSE IMPACTS ON THE ENVIRONMENT

1. **Habitat destruction** There has been extensive land alteration due to agriculture, cattle ranching, urban development and timber production. Logging is common, even within protected areas, to provide firewood and charcoal. Intervals between cuts usually range in the decades but it is often too short a time to allow adequate regeneration. At present, the northern cays (Coco, Romano, Cruz and Paredón Grande) are among the most disturbed areas, mainly as a result of development for tourism. In these cays some species occur only in small numbers and are therefore highly

threatened; examples include Bahama Mockingbird, Thick-billed Vireo and Zapata Sparrow. Habitat disturbance is threatening Piping Plover on the northern cays. In Río Máximo, the most important breeding colony of American Flamingo in the West Indies is also highly threatened, due to anthropic habitat disturbance.

2. **Hunting** Two forms may be recognised: sport hunting, purportedly within the law, and illegal poaching. For sport hunting, there are official regulations regarding species, seasons, places and limits on numbers hunted, but these are frequently disregarded by both hunters and wardens. Birds are regularly killed in numbers exceeding the legal limits and protected species, such as Ruddy Duck and Masked Duck, are sometimes hunted as well. Poachers, on the other hand, persistently violate official restrictions, and use not only guns but also several kinds of traps. The only penalties are confiscation of prey and confiscation of the guns, which happens rarely.

3. **Introduction of exotic species** Both accidental and intentional introductions plague Cuba and other islands in the Antilles. Rats, as well as feral pigs, cats and, more rarely, dogs are now found even in the most remote and virgin forests. These animals, along with Small Indian Mongoose (*Herpestes javanicus*), have undoubtedly had a considerable adverse effect on native bird species, although this has never been documented. Less direct, though perhaps more insidious, has been the impact on habitats of White-tailed Deer (*Odocoileus virginianus*), which has maintained wild populations in many areas for the past 170 years. In addition, over the past two or three decades, Cuba has received more than its share of exotic species (African and Asian bovids and monkeys), along with Wild Boar (*Sus scrofa*). Three different species of monkey have been released on several cays off both the northern and southern coasts, with fragile ecosystems that support several unique bird and reptile subspecies. In the early 1990s catfish (*Clarias* sp.) were introduced with a very negative ecological impact on the indigenous fauna. Several invasive plants have been introduced, including Sicklebush (*Dichrostachys cinerea*), Sweet Acacia (*Vachellia farnesiana*), Australian Pine (*Casuarina equisetifolia*) and Broad-leaved Paperbark (*Melaluca quinquenervia*). The last has had a most detrimental effect on the survival of two relict endemic species: Zapata Rail and Zapata Wren.

4. **Illegal commerce** The species most affected by illegal commerce is Cuban Parrot. Chicks of this species are obtained in the wild by felling nesting trees and rearing them by hand until fully developed. No doubt, hundreds have been smuggled out of the country in a drugged state, simply inside the pockets or luggage of travellers. Other species have recently been detected by Cuban customs, such as Cuban Trogon, Blue-headed Quail-Dove, Cuba Grassquit, Cuban Bullfinch and other native and migratory species. An unfortunate side-effect of tree-felling to obtain parrot chicks is that cavities and trees suitable for cavity excavation by other birds are becoming rare; this is the case for Cuban Screech-Owl, Cuban Pygmy-Owl, Cuban Parakeet, Fernandina's Flicker and other woodpeckers. The illegal trade of bird species also affects Nearctic migrants; the most affected species are Painted Bunting, Indigo Bunting and Rose-breasted Grosbeak.

5. **Chemical pollution** Until the late 1980s, industries on Cuba were developing at a considerable pace. Not surprisingly, sugar refineries and electrical generating plants released a wide range of pollutants that lowered the quality of the air, soil and water. This pollution would have adversely impacted populations of some bird species.

WHERE TO WATCH BIRDS ON CUBA

The best time to observe winter migrants is between October and March when diversity is highest. Summer residents start to arrive in the spring, and for the best photo opportunities we suggest from mid-April to mid-May.

Map of birding sites on Cuba.

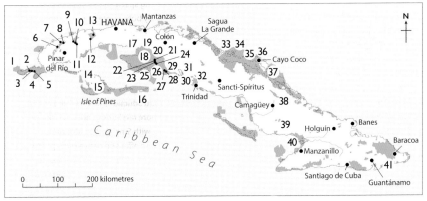

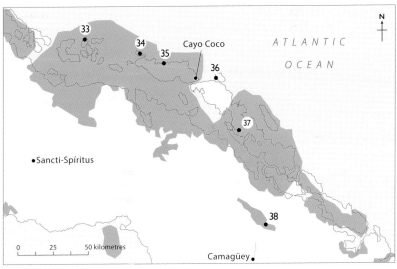

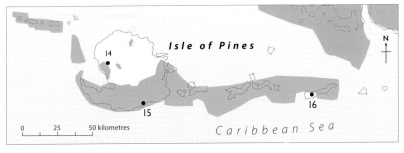

PINAR DEL RÍO PROVINCE
Guanahacabibes Peninsula
Located 380km west of Havana, the peninsula has several vegetation formations dominated by semi-deciduous woodlands. Blue-headed Quail-Dove, Bee Hummingbird, Cuban Parrot, Giant Kingbird and Yellow-headed Warbler can be observed. A total of 16 endemics have been recorded here. Many migratory passerines can also be seen in this area.

Site 1. Las Tumbas
The westernmost tip of the peninsula, with a lighthouse, 59km west of La Bajada. There are woods along the road approaching this site. Birds such as Bee Hummingbird, Cuban Trogon and Cuban Parrot are found in the area. Excellent for migratory birds.

Site 2. El Verraco
This area is 18km west of La Bajada. There are woods with open areas where Bee Hummingbird and Cuban Parrot can be observed.

Site 3. El Veral
This area is 6km west of La Bajada. There are woods with open areas where Gundlach's Hawk and Cuban Green Woodpecker can be seen. This area is about 12km west of La Bajada.

Site 4. María La Gorda
This beach and resort are about 14km from La Bajada. There are woods in the area and about 16 endemics have been recorded, including Blue-headed Quail-Dove, Bee Hummingbird and Cuban Vireo.

Site 5. La Bajada
Ducks and shorebirds can be seen on and around small ponds with mangrove vegetation by the road 800m from the village of La Bajada.

VIÑALES
A beautiful valley with amazing karstic formations called *mogotes* ('haystacks') in the Sierra de los Órganos. It is 26km north of Pinar del Río and 212km west of Havana. A total of 17 endemic species have been recorded here, including Cuban Solitaire and Cuban Grassquit.

Site 6. Sendero Maravillas de Viñales
About 14km west of Viñales town, this area can be good for Gundlach's Hawk, Cuban Tody, Cuban Solitaire, Cuban Grassquit and Cuban Bullfinch.

Site 7. Sierra Ancón
This site is 7km north-west of Rancho San Vicente. Cuban Solitaire and other endemics are present in the area.

Site 8. La Ermita
Approaching La Ermita Hotel, there is a fork between the paved road and a dirt road 200m before reaching the resort. Take the dirt road for 1km to reach a good area for Olive-capped Warbler.

Site 9. Cueva de los Portales, La Güira
Located in Sierra de los Órganos, 150km west of Havana. A total of 18 endemics have been recorded here; among the target species are Scaly-naped Pigeon, Gundlach's Hawk, Giant Kingbird, Cuban Solitaire, Olive-capped Warbler and Cuban Grassquit.

Site 10. Sierra La Güira
About 145km west of Havana. Scaly-naped Pigeon, Gundlach's Hawk, Fernandina's Flicker, Giant Kingbird, Cuban Solitaire and Olive-capped Warbler can be found here.

Site 11. Hacienda Cortina
About 5km from San Diego de los Baños, this site is a garden surrounded by woods. Birds such as Gundlach's Hawk, Fernandina's Flicker and Giant Kingbird can be seen.

Site 12. Soroa
About 87km west of Havana. Sixteen endemics occur here, including Blue-headed Quail-Dove, Grey-fronted Quail-Dove, Cuban Green Woodpecker, Fernandina's Flicker, Cuban Solitaire and Cuban Grassquit.

Site 13. Las Terrazas
Located about 60km west of Havana, 18 endemics are found here. Among the most important are Stygian Owl (a Cuban subspecies), Cuban Trogon, Cuban Green Woodpecker, Fernandina's Flicker and Cuban Grassquit.

ISLE OF PINES

About 100km south of the mainland (Havana). The island has an area of 2,204km². It is mostly flat with a hilly central area that reaches 310m at Sierra de Caballos and Sierra de Casas. Nineteen endemics are recorded here, including Bee Hummingbird and Giant Kingbird.

Site 14. Los Indios
A sandy savanna-like area with many cabbage palms and open vegetation, 34km west of La Fe. Sandhill Crane, Burrowing Owl and Cuban Parrot breed in the area.

Site 15. South of the island
About 20km south of La Fé, the area includes Ciénaga de Lanier (Lanier Swamp) and wooded areas south of Cayo Piedra. Fourteen endemic birds are recorded here.

Site 16. Cayo Largo del Sur
An island 26km long and less than 2km wide, 177km south-east of Havana, from where there are daily flights. Several important species are found here, including Cuban Black Hawk, and Sooty and Bridled Terns can also be seen in summer.

MATANZAS PROVINCE

Site 17. Hato de Jicarita–Río Hatiguanico
This locality lies south of the highway, at the km 101 marker, which is actually 103km from Havana. Drive along the road to Hatibonico River on the right for 7km until it ends at a forest guard station, where you may rent small boats along the Hatiguanico River. Gundlach's Hawk, Zapata Wren, Zapata Sparrow and Red-shouldered Blackbird can be seen, among the 18 endemics recorded.

SITE 18. SANTO TOMÁS
This marsh area was the type locality of Cuba's endemic Zapata Rail, Zapata Wren and Zapata Sparrow. It is about 34km west of Playa Larga. About 112 species have been recorded, including 19 endemics.

Male Cuban Bullfinch.

Site 19. Peralta
Beside the highway between Havana and Santa Clara, Peralta is located at the km 122 marker. The trail runs c. 2.5km alongside swamp woodland, ending in marsh. Fifteen endemics have been recorded here, including Fernandina's Flicker, Zapata Wren, Zapata Sparrow and Red-shouldered Blackbird.

Site 20. La Turba
North of Guamá is an often-productive area of marsh known as La Turba. If coming from the south it is c. 5km beyond Guamá. Zapata Wren, Zapata Sparrow and Red-shouldered Blackbird are among 17 endemics recorded.

Site 21. La Boca
This location is about 12km north of Playa Larga (or 18km south of Jagüey-Grande). It is a tourist centre with scattered trees; several notable species can be observed here, including Bee Hummingbird, Cuban Parakeet, Cuban Parrot, Fernandina's Flicker, Cuban Crow and Cuban Blackbird.

Site 22. Pálpite
Several notable bird species can be observed here, including Bee Hummingbird, Cuban Parrot, Fernandina's Flicker, Cuban Crow, Cuban Oriole, Cuban Blackbird and many migrants. It is 19km south of Jagüey Grande, at the edge of Laguna del Tesoro.

Site 23. Soplillar
This is a small village 5km north-east of Playa Larga. Birds to look for are Bee Hummingbird, Bare-legged Owl, Cuban Pygmy-Owl, Fernandina's Flicker, Cuban Parakeet and Cuban Parrot.

Site 24. Mera
A swamp forest site 1km south-east of Soplillar, Mera has about 16 endemics, including Gray-fronted Quail-Dove, Cuban Nightjar, Bee Hummingbird, Bare-legged Owl and Fernandina's Flicker.

Site 25. La Salina
These abandoned saltpans are about 26km south of Buena Ventura, a small village just 3km west of Playa Larga. This area is covered by mangroves with brackish ponds. American Flamingo, Roseate Spoonbill, shorebirds, egrets and ducks can be seen.

Site 26. El Cenote (Cueva de los Peces)
This limestone-based semi-deciduous forest area is *c*. 18km south-east of Playa Larga on the left-hand side of the road. It is well signposted, 'Cueva de los Peces'. A total of 13 endemics have been recorded, including Blue-headed Quail-Dove and Cuban Trogon.

Site 27 Bermejas
An open area with royal and cabbage palms, bushes and shrubbery 12km north of Playa Girón. The 19 endemics recorded include Cuban Nightjar, Bee Hummingbird, Bare-legged Owl, Cuban Pygmy-Owl, Cuban Trogon, Cuban Tody, Cuban Parakeet, Fernandina's Flicker, Cuban Vireo and Yellow-headed Warbler. More than 100 species have been noted.

Site 28. San Blas
This open area with scattered trees is 14km north of Girón. Fernandina's Flicker and Cuban Parakeet are found here.

Site 29 La Cuchilla
A marshland about 25km north of Girón on the road to Yaguaramas. About 12 endemic species can be observed, including Gundlach's Hawk, Cuban Pygmy-Owl, Cuban Trogon, Cuban Green Woodpecker and Red-shouldered Blackbird. Limpkin and Crested Caracara are also recorded here.

CIENFUEGOS PROVINCE
Site 30. Laguna Guanaroca
This brackish water lagoon of about $2.2km^2$ is located 9km south-east of Cienfuegos city. Several species can be seen, including American Flamingo, ducks and pelicans.

Site 31. Jardin Botánico de Cienfuegos
A 4.5ha botanical garden 23km east of Cienfuegos city. Several endemic birds can be seen, including Cuban Pygmy-Owl, Cuban Tody and Cuban Green Woodpecker.

Site 32. Sierra de Trinidad.
This mountain range is 15km north of Trinidad city. More than 80 species have been recorded, including Black Swift, White-collared Swift, Gundlach's Hawk, Cuban Parrot, Giant Kingbird and Cuban Palm Crow.

VILLA CLARA PROVINCE
Site 33. Cayo Santa María
A cay 39km east-north-east of the town of Yaguajay. It covers $20km^2$ and has a sandy beach 15km long. Several endemic species are present here, including Cuban Black Hawk and Cuban Green Woodpecker, as well as many migrants.

NORTHERN CAYS, CIEGO DE ÁVILA PROVINCE
Site 34. Cayo Guillermo
Situated 34km north-west of Cayo Coco, this is a sandy cay with coastal thickets and lots of *Coccothrinax* sp. palms. Bahama Mockingbird, Mangrove Cuckoo, West Indian Whistling-Duck and American Flamingo occur here.

Site 35. Cayo Coco
This, the second-largest cay on Cuba, is 508km east of Havana and 70km north-west of Morón. It is connected to the mainland by a causeway.

Woodland and wetland habitats cover most of the cay. Over 220 species, including 13 endemics, have been reported. This is an excellent area to find West Indian Whistling-Duck, Key West Quail-Dove, Piping Plover, Thick-billed Vireo, Cuban Gnatcatcher, Oriente Warbler and a subspecies of Zapata Sparrow. There are many waders and large groups of American Flamingo.

Site 36. Cayo Paredón Grande

About 45km east of the Cayo Coco hotel complex, the black-and-yellow lighthouse of Cayo Paredón Grande is recognisable in the distance. The habitat of the cay is coastal thickets and wetlands. More than 100 species have been recorded, including Piping Plover, Cuban Black Hawk, Thick-billed Vireo, Cuban Gnatcatcher, Bahama Mockingbird and Oriente Warbler. Tourist developments have almost totally eliminated the xerophytic habitat.

Site 37. Cayo Romano

The largest of the Cuban cays (connected to the mainland by a rock-filled causeway). Vegetation is similar to that found on Cayo Coco. More than 130 species have been recorded, including 15 endemics, for example Zapata Sparrow and Oriente Warbler.

Juvenile Cuban Black Hawk.

CAMAGÜEY PROVINCE
Site 38. Sierra de Cubitas

The sierra is about 50km north-east of the city of Camagüey. Many birds can be found here, with about 15 endemics including Cuban Pygmy-Owl, Cuban Trogon, Cuban Tody, Cuban Green Woodpecker and Oriente Warbler.

Site 39. Sierra de Najasa

Located about 70km south-east of Camagüey city, this is a protected area of open country with many palm groves and a mixture of semi-deciduous woodland in hill country. About 124 species, including 18 endemics; among the target species are Plain Pigeon, Cuban Parakeet, Giant Kingbird and Cuban Palm Crow.

GRANMA PROVINCE
Site 40. Ciénaga de Birama

A swampy area of around 22km^2 located in the south-east coastal region near the Guacanabayabo Gulf. This swamp is 90km north-west of Manzanillo and south of Jobabo. The Cauto Delta hosts up to 100,000 waterbirds. More than 100 species include Gundlach's Hawk, Fernandina's Flicker, Cuban Parakeet and many migratory ducks.

GUANTÁNAMO PROVINCE
Site 41. Baitiquirí

Located 50km east of the south-coast city of Guantánamo, this xerophytic area is the driest in all of Cuba and is particularly notable for the local subspecies of Zapata Sparrow (*sigmani*). Fewer than 100 species are recorded here, but Cuban Gnatcatcher and Cuban Grassquit are both very common.

INFORMATION FOR VISITING BIRDERS

Planning your trip There are international flights from Europe and America to Havana and Camagüey. Hire cars are available, as well as organised tours, such as KConnell Birds (www.kirkconnellbirds.com). Cuba is a very safe country with friendly, approachable people.

Due to the large number of migrant birds in the avifauna, the optimum time to see the greatest number of species is in autumn, from September to November, or spring, from March to mid-April. The best time to see endemics is from mid-March to May. All the recommended birding sites have wide, well-maintained trails, so walking is easy. We always recommend you have a local guide as the distances are large and specialist knowledge of the birds in the area prevents time lost to searching.

Lodging In some cases it is better to stay in hotels (such as Cayo Coco, Guanahacabibes Peninsula) while in some areas, such as Viñales, Zapata Peninsula and Trinidad, we recommend staying in B&Bs.

Transportation Comfortable buses are provided by Cuban travel agencies, as well as a good taxi service and car rentals (best done in advance from overseas).

Clothing and equipment Off-road birding calls for boots, long trousers, a long-sleeved shirt and hat for protection against the rough bush and the sun. Pack a plastic bag to protect equipment in the rainy months. Always carry water and, if without a guide, a map, compass, GPS tracker and mobile phone.

Water and food Always drink water from bottles and avoid consuming tap water. Many restaurants serve excellent local food.

Security Always lock your car and hide valuable items. Carry cash with you or keep it in a safe box. Never trespass or take photos in military areas or proscribed industrial areas.

Hazards Mosquitoes can be a problem in the rainy season, especially a few weeks after its onset. At this time, adequate repellent must be used when walking through woods near the coast, or on the cays, since mosquitoes are, at times, unbearable even during the day. We recommend wearing long-sleeved shirts and long trousers. During the colder dry season, most of the island is almost a bug-free paradise. Ticks are a problem in pastures with grazing livestock in late winter and spring when the grass is dry. Trousers should be tucked into socks to avoid being infested by small ticks. Spray insect repellant on boots, socks and around the waist. If attacked, undress and wash clothing and remove ticks with tweezers or cover the area with an antihistamine cream. Take care when standing still – if the ground is loosened and grainy, it may be a biting ants' nest. Horse flies are rarely a nuisance, and then only at a few remote beaches. There are several mildly poisonous plants on Cuba, but if you stay on the paths, you should not have any problems. The most common of these noxious plants belongs to the genus *Comocladia* – easily recognisable by its compound, bright green saw-like leaves.

Male Western Spindalis.

HOW TO USE THIS GUIDE

This is the first photographic guide to Cuba, an avifauna that is Neotropical with a large migrant component from the Nearctic region of North America. All species are described, either as regularly occurring species in the main text or as vagrants and rarities in Appendix A.

The systematic order in the species accounts follow the American Ornithologists' Union *Check-list of North American Birds* (AOU, 1998). To update Cuban ornithological information, we have followed *The Birds of Cuba: An Annotated Checklist* (in press.). Systematic arrangement and nomenclature follow supplements to this publication (1999–2017). We also consider four species to be endemic full species that are not presently recognised by the AOU: Cuban Nightjar, Palm Crow, Giant Kingbird and Cuban Bullfinch. Trinomials follow Dickinson and Remsen (2013) and Dickinson and Christidis (2014).

Photographs There is at least one image for each species in the main text, and at least three for each endemic and native resident bird, showing different angles, including flight, dimorphism (differences between male and female) and immature plumages in many. The majority of migratory birds are shown in winter or transitional plumage, with breeding plumage given if the species is regular on Cuba. Three species are illustrated for which photographs were not available: two endemic species, Zapata Rail and Cuban Kite, and one endemic subspecies, Ivory-billed Woodpecker. Yves-Jacques Rey-Millet took the great majority of photographs: native birds on Cuba, and migrants in the Cayman Islands, Dominican Republic, Jamaica, Puerto Rico and North America. Additional photographers are listed at the rear of the book, with major contributions from Nancy Norman, Arturo Kirkconnell Jr. and Bruce Hallett. Habitat photographs were taken by Arturo Kirkconnell.

Range Maps within the text are only shown for those endemic species with a restricted distribution.

SPECIES ACCOUNTS

Species name The common names in English are listed first, followed by the scientific name of each species given in Latin, or Latinised Greek. This name is composed of two words, the first for the genus and the second for the species. The English name is taken from AOU (1998 and supplements). Local Spanish names are taken from Garrido and García Montaña (1975), and Garrido and Kirkconnell (2000).

Taxonomy Taxonomic status is identified as a monotypic species (a taxon that that does not have a recognisable subspecies) or polytypic species (a species that has recognised subspecies); for polytypic species the number of widely accepted subspecies is shown in parentheses.

Description Measurements are taken from Garrido and Kirkconnell (2000). A description is given for each species, along with the most important field marks. Differences between males and females are noted when these are likely to be evident to the field observer. Colour morphs are also described for species in which they are present. The plumages of immature and juvenile birds are also described when these differ markedly from adult plumage, for example in gulls. Non-breeding plumage is usually described first because most migrants occur predominately in this plumage in the Cuban archipelago. Breeding plumage is also described for most species. Shorebirds are regularly seen in early autumn in full, partial breeding or juvenile plumage and full or partial breeding plumage in late spring. A brief description is also given of subspecies or geographical races where appropriate.

Similar species Dominant field characters are described when they usefully distinguish the differences between similar species and the species under discussion. This listing starts with those most similar. Comparison with, and among, migratory species is limited to winter and immature plumages.

Voice Vocalisations of the great majority of birds are translated into words to mimic sounds, based on field observations and recordings. The album *Bird Songs in Cuba,* including the work of G. B. Reynard (Cornell Laboratory of Ornithology, Ithaca, New York), has recordings of 123 species. A copy is useful in the field. Resident birds sing and the phrases are diagnostic; the great majority of migrants do not sing. Most make short diagnostic calls that aid identification, although several warblers sing after arrival or more frequently before departure. Some warblers can even be heard singing in January and February.

Habitats and behaviour The preferred habitats are mentioned for each species (see Vegetation and habitats in the Introduction). This indicates the most likely locations of nesting or foraging birds, both residents and migrants. Behavioural traits contribute to locating and identifying species and include voice, methods of prey capture, tail-flicking and head-bobbing. Understanding habitat use makes it easier to find birds. This involves knowing the substrate they use for nesting (e.g. parrots and woodpeckers are cavity nesters), and also understanding their food sources. Breeding behaviour may include nesting months, courtship, nest shape, clutch size and average brood size. Most landbirds nest in spring and early summer, March–July; some have a longer season from February–September, or breed throughout the year. Large seabirds, such as sulids, tropicbirds and frigatebirds raise only one young during a lengthy nurturing process. Some species, such as Antillean Nighthawk, Gray Kingbird, Black-whiskered Vireo and Least Tern, come to Cuba in summer to breed but are absent in winter.

Range A general description of each species' breeding distribution is given, both inside and outside of Cuba. For North American migrants, both the breeding and wintering ranges in the Americas and the West Indies are given. In all accounts, Cuba refers to the main island of the Republic of Cuba. If a species breeds elsewhere in the Americas, subspecies may be mentioned. Cosmopolitan (pantropical for seabirds breeding within the tropics) is used to describe species that also breed in other parts of the world.

Status Threat status in bold lists species of global conservation concern, as categorised by the IUCN Red list: **Extinct**, **Critically Endangered**, **Endangered**, **Vulnerable** and **Near Threatened**. This is followed by the breeding, migrant and resident status of each species and its distribution, frequency and abundance in preferred habitats. Each part of the archipelago is mentioned: Cuban mainland, Isle of Pines, northern cays and southern cays. Endemic subspecies are included.

Species described as Vagrants and Introduced species are listed separately in Appendices A and B (see pages 342–370).

ABUNDANCE STATUS	
Abundant	Over 20 individuals observed per day in appropriate habitat and season.
Common	Five to 19 individuals observed per day in appropriate habitat and season.
Uncommon	One to five individuals observed in a week in the appropriate habitat and season.
Rare	One to five sightings per year expected in the appropriate habitat and season.
Very rare	One sighting every five years expected in the appropriate habitat and season.
Vagrant	A bird that reaches Cuba by accident, or historically fewer than ten records.
Passage migrant	Migratory bird species that pass through Cuba, usually pausing only briefly.
Summer resident	Migratory bird species that arrive from the south, principally during March, to breed on Cuba, with most departing in September.
Winter visitor	Migratory birds arriving on Cuba from North America in the months from July–to November that spend the winter on Cuba before leaving in February–May.

BIRD TOPOGRAPHY

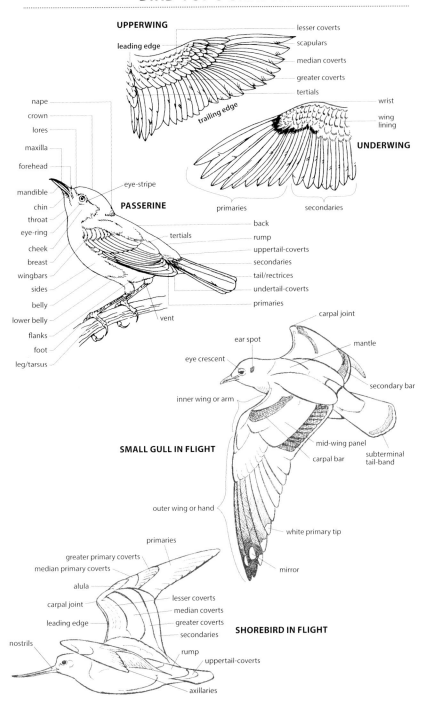

Least Grebe

Tachybaptus dominicus 23–26cm

Adult.

Local name Zaramagullón Chico.
Taxonomy Polytypic (5).
Description Small, greyish-brown grebe with small dark pointed bill, yellow-orange iris, white undertail-coverts; shows white wing-patch in flight. Breeding adults have black crown and throat.
Voice A long trill and a *ping* note.
Similar species Pied-billed Grebe is larger with brownish-grey plumage and black band around large, conical, pale bill.
Habitat and behaviour Wetlands. Breeds mainly May–July, but likely year-round; lays 4–5 bluish, buffy or greenish eggs on nest platform of aquatic vegetation and mud, often floating. Parents cover eggs with plants, causing staining.

Breeding adult pair. Note yellow eye and thin bill.

Range *T. d dominicus* breeds from coastal southern Texas through Middle America to the West Indies, in the Bahamas (except Grand Bahama) and Greater Antilles, where it is common on Cuba and Jamaica, uncommon on Hispaniola and Puerto Rico, absent in the Cayman Islands; mainly sedentary.
Status Common and widespread breeding resident on Cuba, Isle of Pines and northern cays.

Pied-billed Grebe

Podilymbus podiceps 30–38cm

Local name Zaramagullón Grande.
Taxonomy Polytypic (3).
Description Small stocky diver, brownish-grey plumage, large flattened head with blackish crown, pale face, white eye-ring, and fluffy white undertail-coverts. Breeding adult has black chin, throat and thick whitish conical bill with black subterminal band. Migrant adults and resident first-year and non-breeding adults have brownish plumage, pale throat and partial faint ring around bill. Young are greyish-buff with black and white striped face.
Voice Highly vocal with long loud, harsh *ca-cac-cao-wo-kwop-kwop* … over 20 syllables and descending, and pairs have a similar but faster duet; a single low *kuk*; wailing and short barks; chicks and juveniles *peep* constantly.
Similar species Least Grebe is blackish with yellow-orange iris and small pointed bill.
Habitat and behaviour Wetlands. Floats high in water, but responds to threats either by sinking, with only head showing above surface, or diving and surfacing far away. Takes aquatic insects and small fish.

Breeding adult. Note black throat and subterminal band on bill.

Breeds May–February; lays 3–6 whitish eggs; nests on raft of plant matter, either built up from the pond bottom or floating and anchored to adjacent vegetation at the water's edge. Young with red facial skin are often carried on parents' backs.

Non-breeding adult has brownish plumage, pale chin and faint partial ring on bill.

Young have striped heads with orange spot on crown, and red bill and eye-ring.

Range *P. p. antillarum* breeds in the Bahamas, Greater Antilles and Cayman Islands, becoming rare on smaller islands. *P. p. podiceps* breeds from North America to northern South America.
Status *P. p. antillarum* is common and widespread breeding resident on Cuba, Isle of Pines and northern cays; the larger *P. p. podiceps* is rare winter visitor and passage migrant.

Black-capped Petrel

Pterodroma hasitata   35–40cm

Local name Pájaro de las Brujas.
Taxonomy Polytypic (2).
Description Black crown extends below eye, white forehead, brownish-grey upperparts and tail, white rump and uppertail-coverts; white hindneck forms partial collar with white underparts. In flight, blackish leading edge and narrow trailing edge to white underwing, wing bent at 'wrist'.
Voice Silent away from breeding sites.
Similar species None.
Habitat and behaviour Pelagic, and breeds in high mountains. No breeding data from Cuba, with Sierra Maestra the most likely site.

> **Range** Pelagic when not breeding, in Caribbean Sea and Western Atlantic Ocean in Gulf Stream; in the West Indies breeding only confirmed on Hispaniola (formerly Guadeloupe and Dominica), and recently possibly Jamaica.
> **Status** Endangered. Unknown; very rare visitor, no breeding records.

Black-capped Petrel has long wings with white underwing-coverts and distinct black bar.

Audubon's Shearwater

Puffinus lherminieri   30cm

Local name Pampero de Audubon.
Taxonomy Polytypic (4).
Description Small shearwater with entirely blackish-brown upperparts, white underparts; white underwings have conspicuous broad brown margins, undertail-coverts blackish-brown. Short wings with fluttering fast flight with glides close to water's surface.
Voice Silent.
Similar species None.
Habitat and behaviour Pelagic in tropical oceans; breeding confirmed in 2002 under rocks on small northern cay. One breeding record in June.

> **Range** *P. h. lherminieri* breeds locally in the Caribbean Sea on cays and islands off Panama, Nicaragua and Venezuela, and in the West Indies in the Bahamas (common), Cuba, Puerto Rico, Virgin Islands and Lesser Antilles. Pantropical.
> **Status** Previously considered vagrant, now rare breeding resident on Cayo Felipe de Barlovento in the northern cays.

In flight has short wings and long tail. The underside of primaries is mostly white with conspicuous broad smudgy-brown margins and tip.

White-tailed Tropicbird

Phaethon lepturus   81cm, including 30–40cm tail-streamers

Local name Contramaestre.
Taxonomy Polytypic (6).
Description Adult resembles a tern in flight except for diagnostic long central tail-streamers; brilliant white plumage, black streak through eye, orange decurved bill, black on outer primaries and band across inner upperwing-coverts. Juvenile has upperparts heavily barred black and white, yellowish bill and tail-streamers absent from pointed tail; adult plumage by third year.
Similar species Red-billed Tropicbird (vagrant) has red bill, back barred black, in flight black on primaries extends to wingtips and black band on inner primaries is absent.
Voice Constant *cri-et cri-et-cri-et* and *crit-crit crit* heard over long distances; adult screeches and chick screams and hisses in nest-hole when disturbed.
Similar species None.
Habitat and behaviour Pelagic over oceans and around coasts, only coming ashore to breed. Beautiful acrobatic displays with frequent calling in inshore

Juvenile in a small cave in a cliff face. Back and wing-coverts are finely barred black.

waters before flying directly into nest-holes. Plunge-dive for squid and flying fish, chased and robbed by frigatebirds. Breeds March–June; lays one large pinkish-white egg marked with brown in crevices and holes in cliffs or in caves; breeding is prolonged, from laying to fledging is *c.*18 weeks.

Adults in flight show long tail-streamers as well as black on outer primaries and across coverts.

Range Western Atlantic subspecies *P. l. catesbyi* breeds in Bermuda, the Bahamas, Turks and Caicos Islands, Greater Antilles, Cayman Islands, Virgin Islands and Lesser Antilles. Considered threatened in the region where it is in decline on many islands. Pantropical.
Status Rare summer breeding visitor in small colonies in south-east Cuba.

Brown Booby
Sula leucogaster   71–76cm

Adult in flight shows white abdomen, undertail-coverts and underwing.

Local name Pájaro Bobo Prieto.
Taxonomy Polytypic (4).
Description Adult has dark chocolate upperparts, head and neck to mid-breast, sharply defined from white lower breast, belly, undertail- and underwing-coverts, yellow legs and feet. Male has yellow-grey bill with pink and blue facial skin at base; female bill is pinkish-yellow with black spot in front of eye. Juvenile is entirely dull brown, lighter on abdomen, with greyish bill, legs and feet; immature has abdomen mottled with white, brown overall with white tail and rump, dull yellow feet.

Voice Female grunts and honks, and male whistles and croaks on the nest.
Similar species None.
Habitat and behaviour Rocky islands, at sea and inshore waters; plunge-dives for fish and squid. Breeds in colonies on rocky cays, January–May, following prolonged courtship, 1–2 bluish-white eggs laid on the ground in a scrape. Only one young is usually raised (if two eggs hatch, larger sibling pushes the smaller out of the nest), which fledges after 90 days and is fed and protected by both parents for up to nine months.

Adult female.

Immature is brown with mottled white abdomen; it can remain in this plumage for up to two years.

Range Caribbean subspecies *S. l. leucogaster* breeds on islands, with the largest colonies in Puerto Rico, Anguilla, Virgin Islands, Bahamas and Hispaniola. Pantropical; islands off French Guiana and Brazil; Cape Verde, islands off Guinea; Ascension, St Helena, Annobón and throughout the West Indies.
Status Locally common breeding resident on rocky cays off the north and south coasts of Cuba and Isle of Pines.

American White Pelican
Pelecanus erythrorhynchos   150cm

Adults.

Local name Pelícano Blanco.
Taxonomy Monotypic.
Description Very large, heavy white bird with huge yellow-pink bill and orange legs; in flight shows conspicuous black primaries and outer secondaries.
Similar species Brown Pelican is smaller, brownish-grey and plunge-dives for fish.
Habitat and behaviour Coasts and lagoons.

Range Breeds in North America and winters from southern United States to Costa Rica and, rarely, in the West Indies, on Cuba and Puerto Rico; otherwise vagrant.
Status Locally common winter visitor in mainland Cuba, increasing during the last two decades, November–April.

Adults are entirely white except for their black flight feathers.

Brown Pelican

Pelecanus occidentalis 107–137cm

Breeding adult has yellow crown, chestnut nape, silvery-grey plumage and blackish abdomen.

Non-breeding adult has pale yellow head, white neck and dark bill.

First-winter.

Local name Pelícano.
Taxonomy Polytypic (5).
Description Very large, heavy seabird with extremely large long bill and expandable pouch. Non-breeding adult has pale yellow head, white neck, dark bill and iris, grey upperparts. First-winter bird has brownish-grey plumage, whitish underparts and brown iris. Breeding adult has yellow crown, white hindcrown extending as a stripe down foreneck, chestnut hindneck, pale grey coverts and back, dark underparts, white iris.
Voice Silent except at nest.
Similar species American White Pelican is larger, white with dark primaries; does not plunge-dive.
Habitat and behaviour On coasts and brackish lagoons. Flies with slow, deep wingbeats and often glides close to water's surface; flocks fly in line formation; plunge-dives for fish. Breeds in colonies in mangroves year-round; lays 2–4 white eggs.

First-winter bird has brownish-grey crown and nape and pale abdomen.

Range *P. o. occidentalis* breeds in the West Indies. *P. o. carolinensis* breeds on United States coast of South Carolina, on cays off the Atlantic–Gulf–Caribbean to northern South America.
Status *P. o. occidentalis* is a common breeding resident throughout the Cuban archipelago. *P. o. carolinensis* is a common winter visitor and passage migrant in all months mainly in eastern Cuba, with one breeding record.

Neotropic Cormorant

Phalacrocorax brasilianus   66cm

Adult in flight. Note the long tail and dark wings and body.

Local name Corúa de Agua Dulce.
Taxonomy Polytypic (2).
Description Blackish overall with long neck, long bill hooked at tip, small yellowish-orange throat pouch tapering to a point behind bill, dusky supraloral stripe, long tail. When breeding, throat pouch is outlined in white and has white feathers on the side of the head. Immature similar but lighter, especially on underparts.
Voice Low grunts.

Similar species Double-crested Cormorant has orange supraloral stripe and larger orange throat pouch, longer bill and shorter tail.
Habitat and behaviour Freshwater and brackish lagoons, coastal marshes, anthropogenic wetlands and rivers. Breeds in colonies, April–September, building bulky stick nest in trees and bushes growing in water; four pale blue eggs usual.

Adult. Note white V at base of bill and gular pouch, and long tail.

Adult with wings outstretched.

Range *P. b. mexicanus* breeds in southern United States, Mexico to Nicaragua, and in the West Indies in the Bahamas and Cuba.
Status Common breeding resident throughout the Cuban archipelago.

Double-crested Cormorant

Phalacrocorax auritus 74–89cm

Adult in flight.

Adult shows an orange throat pouch and supraloral stripe.

Local name Corúa de Mar.
Taxonomy Polytypic (5).
Description Large, erect, iridescent black seabird with long neck, wings blackish-brown edged black, orange supraloral stripe and throat pouch (seldom visible), green iris and long stiff tail. Breeding adult has two white ear-tufts (not easily visible). Juvenile is similar to adults, but browner above and much paler below, sometimes nearly whitish on the throat and upper breast; throat pouch and mandible are yellowish-orange.
Voice Silent.
Similar species Neotropic Cormorant smaller with smaller yellow throat pouch tapering to point behind smaller bill, dusky supraloral stripe and longer tail. Anhinga has small head, finely pointed bill, silvery coverts and very long tail.
Habitat and behaviour Coasts and wetlands. Buoyant flight with neck kinked. Similar to Neotropic Cormorant and Anhinga: all swim with only head exposed, dive to pursue fish, and adopt a 'wing-spread' position with wings held out to dry. Breeds year-round, in colonies mainly in mangroves; 3–5 bluish-green eggs.

Adult has a dark plumage and orange throat patch.

Juvenile has whitish neck and breast, dark belly, yellowish-orange throat pouch and mandible.

Range *P. a. floridanus* breeds in the United States from North Carolina to Florida and the Gulf coast, and in the West Indies in northern Bahamas and Cuba; fairly sedentary although northern birds disperse south to the Bahamas and Cuba; casual elsewhere in the northern West Indies. *P. a. heuretus* breeds in the Bahamas.
Status *P. a. floridanus* is an abundant breeding resident throughout the Cuban archipelago.

P. a. auritus is a rare winter visitor. *P. a. heuretus* also breeds, but status unknown.

Anhinga

Anhinga anhinga 85cm

Local name Marbella.
Taxonomy Polytypic (2).
Description Adult has small head, long sinuous neck, long fan-shaped white-tipped tail, yellow thin pointed bill. Male is entirely blackish with green iridescence, silvery-white scapulars and secondary upperwing-coverts. Female wings similar but head, neck and breast buffy-pink. Juvenile and immature smaller and browner than female, with reduced white on back and wings. In flight flaps and glides on long pointed wings, showing black underwings and silvery-white on upperwing.
Voice Silent.
Similar species Both species of cormorants lack silvery coverts, have shorter tail and exhibit steady flapping flight.
Habitat and behaviour Wetlands. Swims with head and upper neck above water and dives to pursue fish. Breeds in colonies, March–December, building stick nest in bushes and trees; lays 1–5 bluish-white eggs.

Adult male in flight shows large silvery-white areas on upperwing, and a long neck and tail.

Breeding male shows whitish plumes on head and neck.

Juvenile is paler than female.

Breeding female has blackish-brown head and neck, buffy throat and breast. Also shows silvery-white on scapulars and coverts.

Range *A. a. leucogaster* is mainly sedentary and breeds in south-eastern United States including Florida through the Gulf coast of Middle America to Panama, and the West Indies on Cuba; vagrant elsewhere.
Status Fairly common breeding resident throughout the Cuban archipelago.

Magnificent Frigatebird

Fregata magnificens 94–104cm

Breeding male in flight showing a fully-inflated gular sac.

Immature.

Local name Rabihorcado.
Taxonomy Monotypic.
Description Very large black seabird; long narrow pointed wings bent at the wrist give shallow M silhouette, long deeply forked tail, long bill hooked at tip, small red feet. Male is entirely glossy purple-black, except when red gular sac inflates during courtship and early breeding. Female larger, brownish-black except for white breast and diagonal bronze bars across upperwing-coverts. Juvenile (1–2 years) is brownish-black with white head and abdomen; immature (2–4 years) shows intermediate plumage.
Voice Bill-clattering at nest.
Similar species None.
Habitat and behaviour On coasts, mangrove cays and brackish lagoons and inshore waters. Breeds March–January, when males fly with gular sacs inflated. Colonial nester on offshore cays, building rough stick nests in mangroves or on the ground. Both parents incubate one large white egg for 45–55 days and feed the white fledgling. Feeds on squid and flying fish snatched from the surface, also by kleptoparasitism, robbing Brown Boobies and tropicbirds of fish.

Adult female has a black head and white breast.

Range Pacific and Atlantic coasts of the Americas from Baja California to Ecuador and Galápagos, and from Florida to south Brazil; and locally in the West Indies, with the largest colonies in Barbuda, Cuba, Tobago, Puerto Rico and Virgin Islands, and smaller colonies in the Bahamas, Hispaniola, Cayman Islands, Anguilla, Redonda and St Lucia.
Status Common breeding resident throughout the Cuban archipelago, mainly in the cays.

American Bittern

Botaurus lentiginosus 58–62cm

Local name Guanabá Rojo.
Taxonomy Monotypic.
Description Medium-sized heron. Cryptic plumage of variable rich soft browns with darker streaks on upperparts. Face and throat creamy-buff; black malar stripe continues to sides of breast; underparts buff with dark heavy stripes; bill and legs yellowish-brown. Juvenile lacks malar stripes. In flight shows hunched outline and blackish flight feathers contrast with lighter brown coverts.
Voice Advertising males emit a resonant *oooohnk-A-doonk* for long periods. In flight a nasal *arrk*.
Similar species Immature night-herons have less streaking on underparts and shorter, rounded wings with pale spots.
Habitat and behaviour Mainly freshwater wetlands. Walks slowly and deliberately; points bill upwards and freezes when disturbed.

Adult. Note the cryptic plumage on upperparts, long black malar stripe and bold streaks on underparts.

Adult.

Adult in flight.

Range Breeds from southern Alaska and Canada to Florida and Mexico; winters from southern North America to southern Mexico and the West Indies on Swan Islands, Cuba and the Bahamas, rare elsewhere in the Greater Antilles and Lesser Antilles.
Status Rare winter visitor and passage migrant throughout the Cuban archipelago, August–April.

Least Bittern

Ixobrychus exilis 28–35cm

Local name Garcita.
Taxonomy Polytypic (6).
Description Smallest heron. Adult buffy with dark upperparts, long neck, chestnut nape and wings with bright buffy-yellow patch on coverts, two white lines on back and throat, breast striped white and buff, belly and undertail-coverts white, legs and feet yellowish, bill yellow with dark on maxilla. Male has crown and back black, chocolate-brown in female. Juvenile resembles female but back greyish with greyish-buff patch on coverts, neck and breast buff with darker stripes.
Voice Low rapid three-syllable *hu hu hoo* that descends, harsh *rik-rik-rik* and warning *kek*.
Similar species Green Heron is larger, lacks buffy coverts, has rufous neck and breast.
Habitat and behaviour Freshwater wetlands, including reedbeds, widely though sparsely distributed, and seldom seen. Usually solitary, secretive, freezes with neck pointing skywards; takes fish, aquatic insects, molluscs, toads and lizards by stalking or holding onto cattails and lunging. Breeds April–September; small nest woven into cattails and sedge, 4–5 greenish blue eggs, incubated by female.

Adult female has chocolate-brown upperparts; both sexes have bright buffy-yellow patch on coverts.

Adult male. Note the black crown and back.

Range *I. e. exilis* breeds in western United States, south-east Canada, eastern United States, eastern Mexico to western Costa Rica and Greater Antilles. West Indies population is sedentary. North American birds winter from the Greater Antilles to northern South America.
Status Common breeding resident, winter visitor and passage migrant on Cuba and the Isle of Pines.

Great Blue Heron

Ardea herodias 107–132cm

Local name Garcilote.
Taxonomy Polytypic (5).
Description Very large heavy-bodied heron with long thick neck. Two morphs: grey morph breeding has sturdy yellowish bill with dark maxilla, facial skin blue, grey overall, white head with wide black stripe, long trailing dorsal plumes (absent in non-breeding birds), pale buffy-grey neck with central black-and-white streaking to breast, black 'shoulders', tops of legs chestnut, long legs greyish. In flight shows black flight feathers contrasting with grey coverts. White morph is entirely white, with yellowish or pinkish-olive legs. Juvenile and first-winter bird similar but dark grey overall with all-dark crown and grey-streaked neck.
Voice Usually silent, except alarm call *braak*.
Similar species Great Egret, the only other large white heron, has smaller yellow bill and black legs.
Habitat and behaviour Wetlands. Solitary except on migration and at mixed heron roosts on mangrove cays. Walks slowly or stands motionless to hunt large fish, rodents and reptiles. Flight slow with deep wingbeats and neck folded. Diurnal migrant in small groups (typically 5–6, occasionally up to 17) that arrives on northern cays, September–March. Breeds colonially, February–July, in large stick platform nest in trees or bushes, or cliff ledges; lays 2–3 light bluish-green eggs.

Dark morph in flight.

Dark morph breeding.

Juvenile.

Adult white morph has head plumes and yellowish bill and legs.

Würdermann's Heron shows pale head and upper neck (treated as a subspecies or possibly a hybrid).

Range *A. h. occidentalis* breeds in southern Florida, Yucatán Peninsula, Cuba and islands off north Venezuela. *A. h. herodias* breeds in North America.
Status *A. h. occidentalis* is a common breeding resident throughout the Cuban archipelago. *A. h. herodias* is a common winter visitor and passage migrant throughout the Cuban archipelago.

Great Egret

Ardea alba 89–107cm

Local name Garzón.
Taxonomy Polytypic (4).
Description Large, slender white heron with heavy yellow bill and iris, blackish legs and feet. In breeding plumage, long feathery dorsal plumes extend beyond tail, maxilla becomes dark and lores greenish. Non-breeding adult has yellow bill, lores yellow or grey and legs duller, lacks plumes; juvenile is similar.
Voice Alarmed *krro-aar* on take-off.
Similar species Great Blue Heron white morph is heavier, with thicker bill and pinkish or yellowish legs. All other white herons are smaller: Snowy Egret adult has black bill and yellow feet. Immature Little Blue Heron has blue-grey bill and olive legs and feet. White morph Reddish Egret has bicoloured bill, and juvenile has black bill.
Habitat and behaviour Wetlands, including flooded grassland, agricultural fields, shores. Stalks prey and takes fish, insects, snakes, lizards, small mammals and birds. Breeds colonially, January–September, on large stick platform nest in mangroves (structure less robust than Great Blue Heron); lays 2–4 pale greenish-blue eggs.

Non-breeding adult. It is the only heron with yellow bill and black legs.

Breeding adult has green lores and long feathery dorsal plumes.

First-year.

Adults in flight.

Range *A. a. egretta* breeds throughout the Americas and the West Indies in Jamaica, Cuba and the Bahamas. Northern birds winter south through the breeding range. Cosmopolitan.
Status Common breeding resident, winter visitor and locally very common passage migrant throughout the Cuban archipelago, July–early June.

Snowy Egret

Egretta thula 51–71cm

Local name Garza Real.
Taxonomy Monotypic.
Description Medium-sized heron. Adult entirely white with needle-like black bill, yellow iris, lores and feet, and black legs. Breeding adult develops long plumes on head, breast and back. Juvenile smaller with blackish bill and foreleg, yellow-green feet and stripe on hind leg, black bill and yellow feet develop in second year but yellow leg stripe is retained into second year and, in some cases, in breeding plumage.
Voice Harsh *gaarrh* and low *kroo* in feeding aggregates.
Similar species White herons: juvenile Little Blue Heron has heavier blue-grey bill, entirely olive legs and feet; white morph Reddish Egret is larger with heavy bicoloured bill, and juvenile has black legs and feet; juvenile Great Egret is much larger with heavy yellow bill; Cattle Egret has short stout yellow bill.
Habitat and behaviour Wetlands and flooded grassland. Core species in mixed heron feeding aggregates. Active forager, flying over water in flocks with the leading group 'foot-dragging' yellow feet to attract prey, also stalks prey, waits motionless on a branch and shuffles feet in mud; takes mainly fish, also crabs, aquatic insects, grasshoppers, snakes, toads and snails. Breeds colonially, March–October, in mixed heronries in mangrove, on stick nest platforms, lays 2–4 bluish eggs.

Juvenile in flight shows yellow stripe on hind leg and yellow feet.

Breeding adult displays nuchal and dorsal plumes.

Recently-fledged juvenile shows yellowish-olive feet and blackish bill.

A group of Snowy Egrets foraging.

Range Breeds from northern United States to South America, and the West Indies in the Bahamas, Greater Antilles, Cayman Islands, Virgin Islands and Lesser Antilles. Northern birds winter south throughout the breeding range.
Status Common breeding resident, winter visitor and passage migrant throughout the Cuban archipelago.

Little Blue Heron

Egretta caerulea 56–71cm

Adult. Note purple head and neck, slate-blue body, and bluish-grey bill with black tip.

Local name Garza Azul.
Taxonomy Monotypic.
Description Medium-sized slender heron. Adult has purple head and neck, dark slate-blue body, slightly downturned bluish-grey bill with black tip, dark lores, whitish iris and olive legs. Breeding adult develops long plumes on head, breast and back. Juvenile is white with slate-grey tips to primaries, grey lores, bill and legs as adult; when changing to adult plumage becomes mottled grey-blue and white.
Voice Usually silent; croaks at nest.
Similar species White morph Reddish Egret adult is larger with pink on bill; dark morph adult has rufous head and neck; juvenile has black bill, grey legs and feet (bluish in breeding adult). Juvenile Great Egret is larger with yellow bill and black legs. Cattle Egret is smaller with yellow bill. Juvenile Snowy Egret has fine black bill, yellow lores and front of legs black with yellowish-green stripe on hind leg.
Habitat and behaviour Wetlands and flooded grassland. Takes mainly fish, also crabs, aquatic insects, toads and snails. Breeds October–August, in stick platform nest low in trees (usually mangroves) and usually at the edge of other heron colonies; 3–4 pale greenish-blue eggs.

First-summer transitioning to dark adult plumage.

Juvenile is entirely white with blue-grey base to bill and greenish legs.

Range Breeds in eastern and southern United States to central South America, and the West Indies. North American birds winter throughout the breeding range.
Status Common breeding resident, winter visitor and passage migrant throughout the Cuban archipelago.

Tricolored Heron

Egretta tricolor 61–71cm

Local name Garza de Vientre Blanco.
Taxonomy Polytypic (3).
Description Medium-sized slender heron, blue-grey with white stripe on central neck to grey breast, white belly and underwing-coverts, yellow lores, colour of long bill variable with basal half of mandible yellow and maxilla blackish, darker at tip; yellowish legs. In peak breeding (seen briefly), adult has bright blue lores and bill, red iris, pinkish-red legs, nuchal plumes on crown, long pale chestnut plumes on back and shaggy grey neck feathers. Juvenile has bright rufous neck, mantle and wing coverts gradually becoming mottled with grey in second year.
Voice Harsh *gaarh*, similar to Snowy Egret.
Similar species Adult Little Blue Heron and dark morph Reddish Egret have no white on underparts.
Habitat and behaviour Wetlands. Takes mainly fish, also crabs, insects, toads and snails. Breeds colonially in mixed-species heronries, April–October, in stick platform nest; 3–4 greenish blue eggs.

Breeding adult has blue basal half of bill and lores, white crown plumes and cinnamon-brown dorsal plumes.

Non-breeding adult is duller with yellow lores and legs, and less contrast on bill.

Juvenile. Note rufous on neck and wing-coverts.

Range *E. t. ruficollis* breeds in coastal east and south-eastern United States, to northern South America and the West Indies, in the Bahamas, Greater Antilles, Cayman Islands and, rarely, in the Lesser Antilles. North American birds winter south through the breeding range.
Status Common breeding resident, winter visitor and passage migrant throughout the Cuban archipelago.

Reddish Egret

Egretta rufescens 69–81cm

White morph adult has black or blackish bill.

Local name Garza Roja.
Taxonomy Polytypic (2).
Description Medium-large, thickset heron occurring in white and dark colour morphs. Adult of both morphs has pale iris; long bill is diagnostic with tip half black and basal half dull pink (non-breeding) and bright pink with blue lores and legs (breeding). Dark morph adult is grey with shaggy reddish head, neck and breast. White morph adult is entirely white with dark legs. Juvenile dark morph is grey with paler head and neck, dark lores, bill and legs. Juvenile white morph has blackish bill, pale yellow lores.
Voice Guttural cackle.
Similar species Adult Little Blue Heron is darker with smaller bill, pale blue at base. Juvenile Little Blue Heron has olive legs. Juvenile Great Egret has yellow bill and black legs. Great Egret has yellow bill and black legs. Adult Snowy Egret is slender with thin black bill and yellow feet.
Habitat and behaviour Often solitary on brackish wetlands. Foraging behaviour is diagnostic, dashing, jumping and turning with wings extended, using wing canopy to shade water to attract prey. Breeds in mixed heronries, December–August, in round nest of twigs and roots in mangrove; lays 3–4 greenish blue eggs.

Dark morph adult breeding.

Juvenile or first-year dark morph moulting.

Non-breeding dark morph adult has reddish-brown head and neck and blackish bill.

White morph adult breeding.

Range *E. r. rufescens* breeds in southern United States in Florida and coastal Texas, along the Caribbean coast of Yucatán, and West Indies in Bahamas and Greater Antilles; winters to northern South America, and Greater Antilles and Cayman Islands. It is mainly sedentary with some wandering and post-breeding dispersal.
Status Common breeding resident, winter visitor and passage migrant throughout the Cuban archipelago.

Cattle Egret
Bubulcus ibis 48–64cm

Local name Garza Ganadera.
Taxonomy Polytypic (3).
Description Small, stocky, short-necked white heron with short, thick yellow bill, yellowish-grey legs and dark feet. Breeding adult has red iris, bill and legs (shown briefly), buffy-orange crown, neck, breast and dorsal patches. Non-breeding adult is white. Juvenile is white with dark bill and legs.
Voice Usually silent, alarm call *breeck* and croaks on the nest.
Similar species All are larger. Snowy Egret has fine pointed black bill and yellow feet. Juvenile Little Blue Heron has blue-grey bill and olive legs.
Habitat and behaviour Terrestrial, in savannas, rough and wet pastures, brackish lagoons, anthropogenic areas; stalks insects, lizards, mice and follows livestock to flush prey. Breeds in single-colonies and mixed heronries, April–September, in trees close to water and mangroves; builds stick nest platform, lays 1–5 bluish-white eggs.

Breeding adult has buffy-orange crown, neck, breast and dorsal patches.

Non-breeding adult is entirely white with yellowish bill and yellowish-olive legs.

Juvenile has dark bill and legs.

Range *B. i. ibis* breeds in the Americas and the West Indies (since 1933). Populations from southern Canada and eastern United States migrate to the Greater Antilles in large post-breeding dispersals. This subspecies also breeds in Africa, India and Europe.
Status Abundant breeding resident and passage migrant throughout the Cuban archipelago. It colonised Cuba *c.* 1954.

Green Heron

Butorides virescens 40–48cm

Adult has steel-blue head and back.

Local name Aguaitacaimán.
Taxonomy Polytypic (4).
Description Small, short-legged and short-necked heron with long bill (dark grey maxilla and yellowish mandible), yellow lores, legs and feet. Adult has steel-blue head (raised when alarmed) and back, dark petrol-bluish green wings with coverts edged buffy, rufous neck and breast, central white-streaked band from throat to breast, grey belly and flanks, and short dark tail. Breeding adult shows greenish lores, brighter rufous, grey dorsal plumes and orange legs. Juvenile has dark upperparts, spotted wing-coverts, rufous and white streaks on necks and breast.
Voice Loud *sk-yew* when flushed; warning *kek-kek-kek*.
Similar species Least Bittern is smaller with cinnamon-buff wing-covert patch.
Habitat and behaviour Wetlands. Highly territorial and solitary. Stalks fish, aquatic insects, shrimp, crabs and lizards or perches and waits motionless. Single pairs breed, March–November, in stick platform nest, at 0.3–3.0m above ground, often in Red Mangrove at the edge of lagoons or over water; lays 2–4 greenish-blue eggs.

Adult.

Juvenile has heavily streaked neck and breast; white spots visible on tips of coverts.

Range *B. v. maculata* is confined to the West Indies and is sedentary. *B. v. virescens* breeds from south-eastern Canada and eastern United States to Mexico; winters from southern United States to northern South America and the West Indies.
Status Both subspecies are common throughout the Cuban archipelago; *B. v. maculata* is resident and breeds. *B. v. virescens* is common winter visitor and passage migrant, August–April.

Black-crowned Night-Heron

Nycticorax nycticorax 58–71cm

Local name Guanabá de la Florida.
Taxonomy Polytypic (4).
Description Medium-sized, stocky, short-necked heron with thick bill (shows colour variation according to age), red iris and short yellow legs. Adult has black crown, nape and back, grey coverts without pale edging, white face, neck and throat, red iris and pale grey underparts, blackish bill. Breeding adult has nuchal plumes, greenish or pinkish-red lores and pinkish legs. Juvenile is brownish with large whitish spots on wing-coverts and upperparts, wide cream streaks on buffy underparts, bill with yellowish mandible. In flight feet usually do not extend fully beyond tail.
Voice Call *quork*.
Similar species Yellow-crowned Night-Heron has bill heavier, blunt and curved towards tip, longer legs with feet extending fully beyond tail in flight. Juvenile is greyer with smaller spots on wing-coverts.
Habitat and behaviour Prefers foraging in freshwater and anthropogenic wetlands. Usually solitary, but flocks on migration. Crepuscular, diurnal and nocturnal. Breeds February–September, in small colonies on platform nests in mangrove and woodland; lays 2–5 greenish-blue eggs.

Adult has black crown and back, white underparts and heavy bill.

Juvenile. Note creamy streaks on face and underparts.

Juvenile has large whitish spots on wing-coverts; bill has yellowish mandible.

Range *N. n. hoactli* breeds from North America to South America (south to central Argentina) and the West Indies, in the Bahamas and Greater Antilles; winters throughout the breeding range. Cosmopolitan.
Status Common breeding resident, winter and passage migrant throughout the Cuban archipelago.

Yellow-crowned Night-Heron
Nyctanassa violacea   56–71cm

Adult has black-and-white head and dark grey back.

Local name Guanabá Real.
Taxonomy Polytypic (5).
Description Medium-sized heron with large white head, banded black, dark thick bill, red iris and yellow legs. Neck, mantle and underparts are pale grey, dark grey wing-coverts edged with pale grey. Breeding adult has nuchal and dorsal plumes and yellow wash on white crown. Juvenile has greyish upperparts, thin brown and greyish-cream streaks on underparts, wing-coverts and upperparts finely spotted whitish, bill blackish (in some paler at base). Second-year lacks juvenile spots on coverts and face pattern of adult. Feet extend fully beyond tail in flight.
Voice Loud *kowrk* and repeated *qwok, quwok*.
Similar species Juvenile Black-crowned Night-Heron is browner, stockier, wing-coverts have larger creamy spots and are not edged pale, feet do not usually extend fully beyond the tail in flight.
Habitat and behaviour Brackish and coastal wetlands and woodlands, savannas, open country. Nocturnal. Feeds chiefly on crabs, also fish, insects and young birds. Breeds April–August, as single pairs or small colonies in bulky stick nests from 1–17m above ground in mangroves and coastal woodland and forest; lays 1–4 greenish-blue eggs.

Juvenile. Small whitish spots on wing-coverts and black bill are diagnostic for this species.

Second-year transitioning into adult plumage. Note absence of spots on coverts.

Range *N. v. bancrofti* breeds in coastal regions from Baja California to El Salvador and Nicaragua and the West Indies. *N. v. violacea* breeds in eastern United States and along the Caribbean coast of Mexico and eastern Costa Rica; winters south to Panama and the West Indies.
Status Both subspecies occur throughout the Cuban archipelago. *N. v. bancrofti* is a common breeding resident. *N. v. violacea* is a common winter visitor and passage migrant.

White Ibis

Eudocimus albus 56–71cm

Local name Coco Blanco.
Taxonomy Polytypic (2).
Description Entirely white wading bird with whitish iris, red facial skin, legs and feet, long orange-red decurved bill. In flight plumage is white except for black tips to four primaries. Breeding adult has deeper red facial skin, gular sac and legs, bill black at tip. Juvenile/immature has dark brown back and wings, neck mottled brownish-white, white underparts, orange bill, pinkish-grey legs; in flight shows black primaries and secondaries and terminal band on tail. White adult plumage emerges from first spring.
Voice Low hoarse grunts and a nasal *oohh-oohh*.
Similar species No white herons have long decurved red bill or black tips to primaries. Immature Glossy Ibis is dark overall.
Habitat and behaviour Brackish and freshwater, rice fields. Forages by walking slowly in shallow water,

Adult. Red face and long, pink decurved bill are diagnostic for this species.

probing for invertebrates and vertebrates. Breeds colonially, April–September, in a large platform nest lined with green material in trees and mangroves; 2–5 greenish-white eggs marked with dark brown.

First-year has grey-streaked head and neck, brown upperparts and emerging white feathers.

Adult in flight. Note the entirely white plumage with black tips to primaries.

Juvenile.

Range *E. a. albus* breeds from southern United States to both coasts of Middle and South America to Peru and French Guiana, and the West Indies on Cuba, Hispaniola and Jamaica. Rare in winter in the Bahamas, Cayman Islands and Puerto Rico.
Status Common breeding resident, winter visitor and passage migrant throughout the Cuban archipelago.

Glossy Ibis

Plegadis falcinellus 56–64cm

Breeding adult has iridescent back and wings. Note whitish line at base of dark facial skin.

Local name Coco Prieto.
Taxonomy Monotypic.
Description From distance appears blackish. Non-breeding adult has head and neck finely streaked whitish-brown, iris brown, dark grey lores, pale whitish-blue edge to dark facial skin, iridescent green-bronze on back and wings, long decurved bill and legs grey-pinkish-olive. Breeding adult has chestnut head, neck and body, white on face brighter. Juvenile entirely dark, duller and browner.
Voice Usually silent.
Similar species Juvenile White Ibis has white underparts and red bill; Limpkin is spotted with white on upper body and streaked below, with slower wingbeats.
Habitat and behaviour Wetlands. Forages in shallow water by walking slowly, probing for small invertebrates and vertebrates. Breeds May–September; solitary or in small mixed heron colonies in large shallow nest lined with leaves at moderate height; lays 2–4 dark blue eggs.

Non-breeding adult has brown neck flecked with white.

Range Breeds in eastern North America from Atlantic and Gulf coasts (locally) to Costa Rica and Venezuela, and the West Indies in the Greater Antilles. Resident and non-breeding in the Bahamas. Northern birds winter to the south of the breeding range. Cosmopolitan.
Status Uncommon breeding resident and passage migrant throughout the Cuban archipelago; very large flocks occur locally.

Roseate Spoonbill

Platalea ajaja 66–81cm

Local name Sevilla.
Taxonomy Monotypic.
Description Very large, pink and white wader with very long, flattened grey bill ending in spoon-like tip. Greenish skin on bald head, white neck with black band around nape, whitish mantle, red wing-coverts on pink wings, uppertail ochre with pink or red tail-coverts, dark pink clump of feathers on breast, reddish legs. Colours brighter in the breeding season. Juvenile has white feathered head, neck, back and breast, rest of plumage greyish with pink wash, tail pink, dark legs.
Voice Silent

Adult breeding.

Similar species Only Cuban bird with spoon-shaped bill. Scarlet Ibis (vagrant) and White Ibis are smaller with thin decurved bills. American Flamingo is much larger, adult bright vermilion, with very long neck and legs, in flight shows black primaries and secondaries.
Habitat and behaviour Freshwater and brackish lagoons. Walks slowly, sweeping bill in a wide arc in shallow water to take small fish, insects and weeds. In flight adult appears entirely pink with neck outstretched and slow wingbeats. Breeds August–June; colonial nester in mangroves in bulky stick nest; lays 2–4 white eggs spotted reddish.

Juvenile whitish overall, washed with pale pink on wings. Bill grey, edged pink.

Adult in flight. Note pink plumage and spoon-shaped bill.

Range Breeds in the United States from the Gulf coast of Texas and southern Florida through Middle America to South America, and the West Indies in the Bahamas on Great Inagua and Andros, Cuba and Hispaniola.
Status Common breeding resident throughout the Cuban archipelago.

Wood Stork
Mycteria americana 104cm

Adult in flight displays white plumage with black flight feathers and tail, and long legs.

Local name Cayama.
Taxonomy Monotypic.
Description Very large, white, with dark unfeathered scaly head and neck; long, massive darkish downcurved bill; black on closed wing and tail, blackish legs and pink feet. Immature has downy feathers on head and neck, yellow bill, greyish plumage, and pink legs and feet. In flight entirely white with black primaries, secondaries and tail.
Voice Usually silent.
Similar species None.
Habitat and behaviour Wetlands. Often soars high above ground. Feeds on vertebrates and crustaceans. Breeds January–April; colonial, bulky stick nests in mangrove or other tall trees; lays 3–4 creamy-white eggs.

Adult. Note bald head and neck, huge, slightly downcurved bill, and pink feet.

Juvenile. Note head and neck with down, yellowish bill and pink legs.

Range Southern United States to northern Argentina; Cuba, Dominican Republic.
Status Rare breeding resident on Cuba, Isle of Pines and the larger cays.

American Flamingo

Phoenicopterus ruber   108–122cm

Local name Flamenco.
Taxonomy Monotypic.
Description Adult is very tall, entirely coral-pink, with very long neck and legs; very heavy decurved bill pale at base with large black hooked tip. In flight shows black primaries and secondaries. Adult plumage develops over three years; immature smaller, greyish with fine blackish streaks on back and black-tipped grey bill.
Voice Usually silent, may honk on take-off.
Similar species None.
Habitat and behaviour Wetlands. Filter feeds for algae and very small invertebrates by pumping water through inverted bill. Breeds April–October in colonies of up to 90,000 birds. One egg laid in a mud mound, raised by both parents.

In flight, extends long neck and legs. Pinkish-red overall with black flight feathers.

Group.

Juvenile has black-tipped grey bill, and is greyish with dark streaking on coverts. Adult plumage begins to emerge by third year.

Adult has heavy, black-tipped decurved bill.

Range The Caribbean, breeding on the Yucatán coast, Bonaire, Inagua in southern Bahamas, Caicos Islands, Cuba, and Hispaniola; wanders widely in the region.
Status Common breeding resident throughout the Cuban archipelago.

Black-bellied Whistling-Duck
Dendrocygna autumnalis 46–53cm

Adult in flight. Note large white patch on upperwing.

Local name Yaguasa Barriguiprieta.
Taxonomy Polytypic (2).
Description Goose-like with long legs and neck. No eclipse plumage. Adult has whitish eye-ring, grey face and upper neck; cinnamon breast and back; black belly, rump and tail; pinkish-red or deep orange bill, legs and feet. Large white patch on greater wing-coverts and base of primaries shows as central white stripe on dark wing in flight, head and neck carried low. Immature is dull with grey legs and bill.
Voice Shrill *pe-che-wee-che* whistle.
Similar species West Indian Whistling-Duck has rounded head, black and cream spots on flanks, blackish bill and legs; upperwing-coverts are silver-grey in flight. Fulvous Whistling-Duck is smaller, cinnamon overall including barring on dark back; in flight dark wings are unmarked with conspicuous white rump.
Habitat and behaviour Wetlands. No breeding data.

Adult pair.

Adult has whitish eye-ring, black abdomen, and pink bill and legs.

Range *D. a. fulgens* breeds in southern Arizona, Texas, Middle America to Panama and Cuba.
Status Very rare winter visitor on mainland Cuba.

West Indian Whistling-Duck

Dendrocygna arborea 48–56cm

Local name Yaguasa, Cuba Libre.
Taxonomy Monotypic.
Description Long-legged, long-necked, upright duck, similar to a goose. No eclipse plumage. Adult has chestnut-brown forecrown, crested blackish hindcrown and nape; wing-coverts edged buff, flanks with large creamy-white central area on each dark feather. Lower face, throat and upper neck grey-buff, lower neck and breast tawny-chestnut with black streaks; whitish belly and undertail-coverts spotted with black, mid-back, rump and tail blackish; bill, legs and feet greyish-black. In flight head and tail are held low, legs project beyond tail; shows silvery patch on inner primaries and greater primary coverts, unmarked underwings. Juvenile smaller, duller with less spotting.
Voice Far-carrying three-, four- or five-syllable whistle- *tsssee-tsssee-tssee-seee* (rising)-*tsweer*.
Similar species Black-bellied Whistling-Duck has cinnamon back and breast, black belly, red bill, legs and feet, and broad white band on upperwing in flight. Smaller Fulvous Whistling-Duck is cinnamon overall with white flashes on flanks; in flight shows black tail with white uppertail-coverts and dark unmarked wings.
Habitat and behaviour Wetlands. Usually in very noisy flocks. Roosts on trees, among weeds or on top of Red Mangrove roots. In Cayo Coco resorts can be found under bungalows. Take fruits (Royal Palm), bulbs and grass seeds; frequently upends for weeds with small invertebrates attached. Breeds in all months, peaks April–January; territorial and aggressive. Nests in holes of Red Mangrove and Buttonwood trees, bromeliads, tall reedbeds or grassland near water; lays up to 16 white eggs, young in large crèches supervised by several adults.

Adult in flight. Note silvery inner primaries and greater primary coverts.

Adult. Dark side feathers with whitish centres are diagnostic for this species.

Adult with young.

Range West Indian endemic species; its range has contracted with significant populations now only in the Bahamas, Turks and Caicos Islands, Cuba, Dominican Republic, Cayman Islands, and Antigua and Barbuda.
Status Vulnerable globally and nationally threatened. Widely distributed, mainly in central Cuba, and fairly common breeding resident throughout the Cuban archipelago. Main threats are habitat loss and illegal hunting.

Fulvous Whistling-Duck

Dendrocygna bicolor 46–51cm

Local name Yaguasin.
Taxonomy Monotypic.
Description Large goose-like duck. No eclipse plumage. Adult has face, breast and underparts unmarked cinnamon; dark brown back barred tawny; defined buffy-white, elongated flank feathers. Bill, legs and feet dark grey. In flight head held low and legs extend beyond tail, shows blackish unmarked wings, white uppertail-coverts contrasting with dark tail. Immature similar but duller.
Voice Two-syllable whistle *wu-c*.
Similar species West Indian Whistling-Duck has rounded head, black and cream pattern on sides; in flight upperwing-coverts show silvery patch. Black-bellied Whistling-Duck has black abdomen, whitish eye-ring, pinkish-red bill, legs and feet; in flight shows large white band along central upperwing.
Habitat and behaviour Wetlands. Usually in small flocks. Breeds April–December, lays 11–18 pale yellow or whitish eggs in rice plants, grass beds and reeds.

Adult in flight. Note dark unmarked wing, white uppertail-coverts and dark tail.

Adults show tawny barring on dark back and buffy-white elongated side feathers.

Range Breeds in North America from southern California and United States Gulf coast to southern Florida, both coasts of Middle America to northern Argentina, and West Indies.
Status Common, but locally distributed, breeding resident on Cuba, Isle of Pines, Cayo Sabinal and Cayo Coco. It was first reported in 1943.

Wood Duck

Aix sponsa 45–51cm

Local name Pato Huyuyo.
Taxonomy Monotypic.
Description Male has iridescent green crest, white 'straps' on face and neck, orange-and-black bill with extensive red at base, rufous-plum breast and bright yellow-buff sides and flanks. Female and eclipse male are crested with small grey bill, broad white asymmetrical eye-ring, brownish-grey plumage with underparts lighter and spotted.
Voice Female take-off call *whoo-eek*.
Similar species None.
Habitat and behaviour Wetlands, usually in pairs or small groups. Breeds June–October; lays 8–14 white eggs in tree cavities and dead palms.

Eclipse male. Note white throat, neck ring and cheek stripe, as well as red eye and base of bill.

Female has large oval area around dark eye.

Breeding male has iridescent green crown and long crest.

Range Breeds in North America south to Gulf coast, southern Florida, Mexico and the West Indies on Cuba and Bahamas.
Status Uncommon breeding resident and winter visitor on Cuba, Isles of Pines and northern cays.

Gadwall

Mareca strepera 46–57cm

Local name Pato Gris.
Taxonomy Polytypic (2).
Description Dabbling duck. Adults and juvenile have steep forehead. Female has thin bill with grey maxilla and orange sides, plumage speckled and mottled brown; in flight white underwing-coverts and all plumages show diagnostic white inner secondaries (speculum), often visible at rest. Breeding male has brown head, slaty to black bill, grey overall with speckled breast, black rump and tail coverts, long silver tertials and cinnamon-edged scapulars. In eclipse plumage, male has similar bill colour to female.
Voice Male gives high-pitched whistle and low croaking *gua*, female a quacking *gaa-gaa-gaa*.
Similar species Mallard has violet-blue speculum; male has yellow bill and female has dark patch in centre of yellow bill. Female American Wigeon has blue bill, green speculum and white on forewing.
Habitat and behaviour Wetlands.

Male. Note greyish-brown head with steep forehead, light face, grey body with cinnamon-brown scapulars, blackish bill and black rump patch.

Female is speckled brown overall; bill has orange sides and dark centre. Only duck with white secondaries.

Range *M. s. strepera* breeds in North America, north and central Eurasia; North American populations winter to the Gulf coast and Florida, Mexico, and rarely in the West Indies in the Bahamas, Turks and Caicos Islands and Cuba; vagrant elsewhere in the Antilles.
Status Very rare winter visitor and passage migrant mainly on Cuba and northern cays, October–May.

American Wigeon

Mareca americana 46–56cm

Local name Pato Lavanco.
Taxonomy Monotypic.
Description Male has whitish crown and iridescent wide green band from eye to nape, small pale bluish-grey bill with black tip, pinkish-brown breast and sides; white flank patch contrasts with black rear and pointed tail. Eclipse male and adult female have grey speckled head and tan breast and sides. In flight both sexes show green speculum; male has white upperwing-coverts (forewing), brown in female edged white on greater coverts, shows as white bar across dark wing. Pale-variant females also show white upperwing-coverts.
Voice *Quack* or three whistles.
Similar species None.
Habitat and behaviour Wetlands, mainly brackish lagoons, in flocks of 10–200.

Adult male. Note white crown and flank patch, and green iridescent area from eye to nape.

Adult female.

Range Breeds in Alaska, Canada and northern United States; winters from western North America south through Central America to northern Colombia, and the West Indies, where common on Cuba and Hispaniola, uncommon in the Bahamas and rest of Greater Antilles, rare in Lesser Antilles.
Status Common winter visitor and passage migrant to Cuba, Isle of Pines and northern cays, August–May.

Mallard

Anas platyrhynchos 51–71cm

Local name Pato Inglés.
Taxonomy Polytypic (3).
Description Male has iridescent green head and neck, white collar and chestnut breast, yellow bill and legs. Tail white with curled black tail-coverts. Female and eclipse male are mottled brown and yellowish-buff; female has orange bill with black markings in the centre; tail has white margins; eclipse male and juvenile have olive-yellow bill. In flight all plumages show violet-blue speculum bordered with white, tail and underwing white.
Voice Loud *quack*.
Similar species Male Northern Shoveler smaller, bill longer and spatulate, white breast and chestnut sides, and shows green speculum and blue forewing in flight.
Habitat and behaviour Wetlands.

Male. Note iridescent green head and neck, white collar and black rump.

Female has dark crown and eye-stripe. Both sexes show violet-blue speculum, bordered with white.

Range *A. p. platyrhynchos* breeds in North America; winters in North America south to the Gulf coast, central Mexico, casual reports in the West Indies in Bahamas (Andros, New Providence), western Cuba, Cayman Islands, Puerto Rico, the Virgin Islands (St Croix).
Status Very rare winter visitor and passage migrant on Cuba and Isle of Pines, September–May.

Blue-winged Teal

Spatula discors 38–40cm

Local name Pato de la Florida.
Taxonomy Monotypic.
Description Small ashy-brown dabbling duck. Adult male densely spotted with black, with conspicuous white crescent before eye, long blackish bill, white flank patch, black undertail-coverts, yellow-orange legs and feet. Adult female and eclipse male speckled grey-brown feathers edged buffy, black eye-line; adult female has white broken eye-ring, throat and lores (spot at base of bill). In flight both sexes show light blue upperwing-coverts, and green speculum. Juvenile similar to female.
Voice Females *quack* loudly.
Similar species Green-winged Teal female has black eye-line, lacks white spot on lores; both sexes show grey upperwing-coverts in flight. Northern Shoveler female has large spatulate bill.
Habitat and behaviour Wetlands.

Female. Note dark eye-line, broken white eye-ring and whitish spot at base of bill.

Male in flight shows blue upperwing-coverts and green speculum.

Adult male. The white facial crescent is diagnostic.

Range Breeds in southern Canada and United States, locally on the Gulf coast and Florida; winters from the southern United States to the West Indies and central Argentina; most common migrant duck in Bahamas and Greater Antilles.
Status Abundant winter visitor and passage migrant throughout the Cuban archipelago, August–June. Up to one million present in winter.

Northern Shoveler

Spatula clypeata 43–53cm

Local name Pato Cuchareta.
Taxonomy Monotypic.
Description Large dabbling duck with very wide spatulate bill. Male has iridescent blackish-green head and neck, white breast, chestnut underparts with large white patch before black tail, orange legs and black bill. First-winter male resembles adult female, may have white on face; black bill; green shows on head by February and chestnut on sides emerges by April. Adult female and eclipse male have greyish-brown plumage broadly edged with buff; female has dark yellowish-black bill edged orange (can appear bright yellow in the field). In flight male has light blue upperwing-coverts, grey in female; both have green speculum, extensive white on underwing-coverts, and spatulate bill emphasises size of head.
Voice Male cackles; female weak *quack*.
Similar species Only duck with large spatulate bill. Mallard has chestnut chest, white abdomen and shorter yellow bill. In flight female Blue-winged Teal is smaller with similar wing pattern but bill much smaller.
Habitat and behaviour Wetlands. Groups smaller than ten usual.

Adult female. Note spatulate bill, which is dark on top with orange edges.

Male in autumn has black bill and green plumage developing on head. First-winter bird appears similar but bill is edged yellow.

Adult male.

Range Northern and central North America; winters from the United States to northern South America and the West Indies, where common on Cuba, otherwise uncommon to rare. Palearctic.

Status Common winter visitor and passage migrant on Cuba, Isle of Pines, Cayo Coco and Cayo Sabinal, July–May.

White-cheeked Pintail

Anas bahamensis 38–48cm

Local name Pato de Bahamas.
Taxonomy Polytypic (3).
Description Elegant slender duck with light and dark brown plumage, long whitish-buff tail, diagnostic white cheeks and throat, and red at base of long grey bill. In flight shows green speculum bordered by buff. Female duller than male.
Voice Males squeak; females *quack*.
Similar species None.
Habitat and behaviour Wetlands. Breeds May–August, nest of grasses and aquatic vegetation on the ground near water with 6–12 cream-coloured eggs.

Adult has long bill, red at base; rounded, soft brown head with white cheeks; and pointed white tail.

Male (right) has more pronounced red at base of bill and long tail is white, whereas female (left) has less red on bill and tail is buffy.

Range *A. b. bahamensis* breeds in coastal South America and the West Indies in the Bahamas, Greater Antilles and some islands in the Lesser Antilles; vagrant in Jamaica.
Status Uncommon, but local, breeding resident on Cuba and northern cays.

Northern Pintail

Anas acuta Male 72cm • Female 55cm

Local name Pato Pescuecilargo.
Taxonomy Monotypic.
Description Slender dabbling duck with rounded head, long neck, wings and pointed tail. Male has dark brown head, nape and throat, white stripe from nape to white breast, bright blue-grey sides to bill, silver-grey flanks, white flank patch before black rear, tail with long black central feathers. Female is mottled light brown with dark bill and reddish edges to feathers on upperparts with shorter pointed brown tail. Eclipse male has greyish plumage, pale brownish head and white neck. Both sexes have long scapulars (longer in males). In flight both show long pointed wings and tail, male has dark upperwings with iridescent green speculum (appears blackish), bordered in front by buff and behind by bold white line. Female has brown speculum with broad white trailing edge.
Voice Male, a rarely heard double whistle. Female, a low croak.
Similar species West Indian Whistling-Duck is heavier with darker wings and different flight profile, head, neck, and legs carried low, slower wingbeats. Female Gadwall has white speculum bordered with black, and grey bill with orange sides.
Habitat and behaviour Wetlands, often in small flocks.

Adult male. Note chestnut head, white stripe from nape to breast, blue-grey bill, and long pointed tail.

Adult female is brown with rounded head and long pointed tail.

Range Breeds in northern North America; winters south to Central America and northern South America, Bermuda, and the West Indies, where common on Cuba, uncommon to rare in the Bahamas, Hispaniola, Puerto Rico and the Cayman Islands. Palearctic.

Status Rare winter visitor and passage migrant on Cuba, Isle of Pines, Cayo Coco and Cayo Sabinal, September–April. This species has been in decline in the last two decades.

Green-winged Teal

Anas crecca 33–39cm

Local name Pato Serrano.
Taxonomy Polytypic (2).
Description Small brown dabbling duck with round head and small black bill. Male has chestnut head and neck, iridescent green band from eye to nape outlined with white, greyish back with long scapulars, buffy breast spotted cinnamon, flanks finely barred grey, white vertical bar in front of wing, white belly, dark rump and tail with a prominent buffy area under tail. Eclipse male and adult female mottled greyish-brown with buffy edges to feathers, pale buff on undertail-coverts. Female has dark crown and eye-line and faint eye-ring. In flight both sexes show dark upperwing-coverts separated from green speculum by buff bar, and white trailing edge; white centre on underwings.
Voice Usually silent.
Similar species Blue-winged Teal female has white lores and both sexes show blue upperwing-coverts in flight.
Habitat and behaviour Wetlands. Usually solitary.

Adult male. Note chestnut head, iridescent green crescent to nape and white bar in front of wing.

Adult female has a blackish crown and eye-line. In flight, females show a green speculum with an inner brown bar on the upperwing.

Range *A. c. carolinensis* breeds in northern North America; winters throughout North America to southern Florida, rarely to Central America and the West Indies, where uncommon on Cuba and rare in the rest of Greater Antilles and Cayman Islands.

Status Uncommon, but local, winter visitor and passage migrant on Cuba, Isle of Pines, and northern cays, August–April.

Canvasback

Aythya valisineria 56 cm

Male has white body, reddish-brown head and neck, and black breast, rump and tail.

Female has pale brownish-grey back and sides, and brown head and neck.

Local name Pato Lomiblanco.
Taxonomy Monotypic.
Description Diving duck with large triangular head and thick neck. Male breeding has dark chestnut-brown head, black breast, lower back, rump and tail, rest of body whitish. Eclipse male similar but body brownish-grey. Breeding female has pale brownish-grey back and sides, with brown neck, chest, rump, breast and tail. Both sexes have long blackish bill and sloping forehead. In flight wings lack contrasting pattern, upperwing extensively white in male and greyish in female.

Voice Usually silent.
Similar species Male Redhead has grey back, short, tricolored bill, rounder head and grey trailing edge to wing.
Habitat and behaviour Wetlands. Patters when taking off from water; flight very fast. Migrating flocks fly in irregular V-formations or in lines.

Range Breeds in western and central North America, wintering to southern Mexico, Gulf coast and Florida.
Status Very rare winter visitor, November–April.

Redhead

Aythya americana 51 cm

Male has red head and neck, black breast and rump, silver upperparts, and blue bill with black tip.

Female is brown with pale face and throat, slate-grey bill with black tip and whitish eye-ring.

Local name Pato Cabecirrojo.
Taxonomy Monotypic.
Description Medium-sized diving duck with steeply rising forehead and rounded head. Male is grey with reddish-brown head, black breast, rump and tail-coverts, yellow eye, white belly. Female tawny brown with greyish abdomen, and pale buffy eye-ring and face around the bill. Both sexes have slate-grey bill with a white ring and black tip, brighter in male. In flight pale grey speculum with narrow white borders and dark coverts. Immature resembles female.
Voice Usually silent.
Similar species Male Canvasback has whitish back, long blackish bill and sloping forehead.
Habitat and behaviour Wetlands. Flight fast and strong; runs over water to become airborne.

Range Western and south-eastern Canada and north-western and north-eastern United States, wintering south to Guatemala, Bahamas and Jamaica.
Status Very rare winter visitor, November–March.

Ring-necked Duck

Aythya collaris 40–46cm

Local name Pato Cabezón.
Taxonomy Monotypic.
Description Diving duck with peaked crown. Male has iridescent violet-black head, neck and back; bill slate-grey, edged white, with narrow whitish line at base, white subterminal ring and wide black tip; black breast with white vertical bar in front of wing, flanks finely barred pale grey, and rest of underparts white. Female is brownish overall with darker crown, back and breast; white eye-ring may have postocular line, face mainly grey, whitish patch at base of dark grey black-tipped bill with thin white subterminal ring. Eclipse male resembles breeding male but duller, with brownish-grey flanks. In flight both sexes show grey secondaries contrasting with dark upperwing-coverts and primaries.
Voice Silent.
Similar species Lesser Scaup has no white on bill, back is greyish, flanks appear white in field; female lacks white eye-ring and postocular line. In flight both show white secondaries with dark wingtips.
Habitat and behaviour Wetlands, usually in large compact flocks.

Female. White eye-ring, postocular line and ring on bill are diagnostic in this species.

Male. Note slate-grey bill with black tip and white subterminal ring.

Range Breeds in northern North America; winters in the United States and Central America to Panama and in the West Indies to Grenada, where common in the northern Bahamas and Cuba, uncommon in Jamaica and variable numbers in the Cayman Islands; rare or vagrant elsewhere.
Status Common winter visitor and passage migrant to Cuba and Isle of Pines, Cayo Coco, Cayo Sabinal, August–April.

Lesser Scaup

Aythya affinis   38–46cm

Local name Pato Morisco.
Taxonomy Monotypic.
Description Diving duck with peaked hindcrown and slate-grey bill with black terminal spot, yellow iris. Male appears black and white in the field; iridescent black head and neck, breast and tail, back speckled grey, sides are speckled pale greyish-white. Female has broad white patch at base of bill, dark brown head, breast and back, paler on flanks. Eclipse male has brownish wash overall, no white at base of bill. In flight both show white secondaries with dark trailing edge.
Voice Silent.
Similar species Ring-necked Duck male has white subterminal ring around blue bill, white bar in front of wing and black back; female has white eye-ring and postocular line. In flight both show pale grey secondaries.
Habitat and behaviour Freshwater wetlands, usually in large flocks.

Male has speckled grey back, white sides, yellow eye in black face and slate-grey bill with black tip.

Breeding female has white patch at base of bill that becomes less distinct during winter.

Range Breeds in northern North America; winters from North America to northern South America and the West Indies where fairly common in northern Bahamas, Cuba, Jamaica and Cayman Islands; uncommon elsewhere in Greater Antilles.
Status Locally common winter visitor and passage migrant throughout the Cuban archipelago, August–April.

DUCKS

Hooded Merganser

Lophodytes cucullatus   48cm

Local name Pato de Cresta.
Taxonomy Monotypic.
Description Male has black head with a distinct white patch, black neck and back, white breast with two black bars laterally, rufous sides. Female is brown, darker on back. Both sexes have yellow eyes, thin short black bill with yellowish-orange mandible. Crest is rounded, erected to form fan-shaped white patch in males, rusty orange and bushy in females. In flight crest is flattened; both sexes show narrow wings and long tail, with small white speculum.
Voice Croaking notes.
Similar species Red-breasted Merganser is larger with long red bill and distinct large white patches across inner wing in flight.
Habitat and behaviour Wetlands. Flies fast with shallow wingbeats. Crest is raised and lowered during courtship or antagonistic interactions.

Male (right) has black crested head with distinctive white patch. Female (left) has brownish crest.

Adult male. Note short thin bill; both sexes have yellow eyes.

Range Breeds in North America south to the Mississippi Valley; winters south to northern Mexico and the northern Bahamas and Greater Antilles.

Status Rare winter visitor and passage migrant to coasts of Cuba and Cayo Coco, November–May.

Red-breasted Merganser

Mergus serrator 51–64cm

Local name Pato Serrucho.
Taxonomy Monotypic.
Description Diving duck. Adults have shaggy crest and thin red bill, hooked at tip. Male has iridescent green head, broad white neck-ring, black back with extensive white on wing-coverts, rufous breast streaked with black, and grey sides. Female has brownish-rufous crest, whitish chin, grey back and sides. Eclipse male similar to female. In flight body is stretched horizontally; male shows dark unmarked head, white neck-ring, white inner wing crossed by two black bars, reduced in female; both sexes show white on underwing.
Voice Silent.
Similar species Hooded Merganser is smaller with dark bill and longer tail; female has rounded crest; brown back, sides and flanks, white speculum less prominent; wingbeats are faster.
Habitat and behaviour Inshore waters, open sea near coasts, brackish water wetlands. Observed in scattered flocks. Flight low. Females and immatures more common than males.

Female has reddish-brown head, reddish bill and grey breast. White inner wing is visible in flight.

Male has green head with crest, white collar and long, reddish hooked bill.

Range Breeds in northern North America; winters in North America including the Gulf coast to Florida, and very rarely in the West Indies in the northern Bahamas, Greater Antilles and Cayman Islands. Cosmopolitan.

Status Locally common, and increasing, winter visitor on Cuba and the northern cays, October–May.

Masked Duck

Nomonyx dominicus 30–36cm

Local name Pato Agostero.
Taxonomy Monotypic.
Description Small stiff-tailed duck with flat crown. Breeding male has black face, cinnamon-rufous plumage with black spotting on back and flanks, and black tip to blue bill. Female, eclipse and winter male and juvenile have plumage barred brownish-grey, black cap and two black stripes across buffy face, bill grey. Tail often raised. In winter, male resembles female with darker face stripes. In flight adults show diagnostic white curved patch on inner upperwing.
Voice Mostly silent.
Similar species Ruddy Duck has dark wings in flight; male has black head with white cheeks and entirely blue bill; female and immature have single dark line across face.
Habitat and behaviour Freshwater wetlands. Secretive. Breeds June–September, laying 8–18 buffy white or cream-coloured eggs.

Female has dark crown and two horizontal stripes across face.

Breeding male. Note blue bill with black tip and black mask. Typically found in dense vegetation.

Range Resident locally in North America on the United States Gulf coast, Middle America to South America, and the West Indies in the Bahamas (rare), Greater Antilles and Lesser Antilles.
Status Rare breeding resident on Cuba and Isle of Pines. Considered threatened locally.

Ruddy Duck

Oxyura jamaicensis 35–43cm

Local name Pato Chorizo.
Taxonomy Polytypic (3).
Description Small, compact, stiff-tailed diving duck. Breeding male is reddish-brown with black cap, white cheeks and blue bill. In eclipse similar to female but cheeks remain white, underparts browner and heavily barred. Adult female and first-winter birds have greyish-brown upperparts, buffy-white cheeks crossed by a single dark line, dark bill. In flight both sexes show dark, unmarked upperwing.
Voice Mostly silent.
Similar species See Masked Duck.
Habitat and behaviour Freshwater wetlands. Winter visitors usually gather in flocks. Breeds June–August; lays up to six whitish eggs in dense aquatic vegetation.

Adult female (seen here) and juvenile have single dark stripe across cheeks.

Breeding male has white cheeks and pale blue bill. Reddish-brown overall and tail often held vertically.

Non-breeding male is brown with blackish crown and white cheeks.

Range *O. j. jamaicensis* breeds in the West Indies in the Bahamas, Caicos Islands and Greater Antilles. North American *O. j. rubida* winters south to southern Mexico; also recorded on Cuba.
Status *O. j. jamaicensis* is a rare breeding resident; *O. j. rubida* is a locally common winter visitor to Cuba, Isle of Pines, Cayo Coco and Cayo Sabinal.

Osprey

Pandion haliaetus 53–61cm

Local name Guincho.
Taxonomy Polytypic (4).
Description Sexes alike, female larger and heavier. Migrant *P. h. carolinensis* has dark brown upperparts, white head with dark stripes on crown, and wide band from eye to hindneck, white neck and underparts with streaks on breast varying in density depending on sex (darker on females sometimes forming a necklace). Caribbean *P. h. ridgwayi* has very pale head and faint mark on ear-coverts, white breast, duller brown upperparts. In flight wings are long and narrow, angled at carpal joints to form a shallow V, white underparts and underwing-coverts contrast with black carpal patches, barred primaries, secondaries and long tail. Immature has mantle and wing-coverts edged pale.
Similar species None.
Voice Far carrying *s-ee-ew* repeated, alarm *kui kui kui*, and series of whistles.
Habitat and behaviour Fringing reefs, marine sounds, fresh- and saltwater wetlands. Makes spectacular plunge dives, also takes fish near surface diving feet-first into water. Diet mainly fish. Breeds October–May; nest is accumulation of sticks and debris near water on lighthouses and tops of mangroves; lays three eggs, creamy to yellowish, spotted and blotched with chestnut-red to dark brown.

In flight adults display barred flight feathers and black carpal patches on underwing.

Caribbean subspecies *ridgwayi* has almost entirely white head and white breast.

Migrant subspecies *carolinensis* has blackish stripe from eye to nape; blackish feathers on breast form indistinct breast-band.

Range *P. h. ridgwayi* breeds in the Caribbean including Cuba, the southern Bahamas and Belize. *P. h. carolinensis* breeds in North America and winters to South America and the West Indies though usually south of range of sedentary, paler Caribbean race. Cosmopolitan.
Status *P. h. ridgwayi* is a rare and local breeding resident on offshore cays and, rarely, on the mainland in the Zapata Peninsula. *P. h. carolinensis* is a common winter visitor and passage migrant throughout the Cuban archipelago; has bred once on Cuba.

Turkey Vulture
Cathartes aura 68–80cm

Adult has blackish plumage and red legs. Head naked with reddish skin, often reddish-purple in Cuban birds.

Local name Tiñosa.
Taxonomy Polytypic (6).
Description Large bird with broad wings and small head. Adult is dark brown, red bill pale at tip and hooked, head and neck have purplish-red bare skin, and reddish feet. Juvenile is dull brown with grey skin on head and neck. In flight wing-coverts and flight feathers are silvery, tail is long and rounded.
Voice Silent.
Similar species Black Vulture has dark head and short square tail. In flight shows a white patch on wingtips.
Habitat and behaviour Over open areas and marshes. Large wing area allows soaring and tilting flight using thermals; wings held above horizontal in a wide V, outer primaries are widely spaced. Forages using highly developed sense of smell; almost exclusively a scavenger, feeding on carrion of rats, mice, snakes and birds, often at roadkills and on shores. Breeds January–June, nest on the ground in rocky cliffs, caves, sugar cane fields or on forest floor; lays 1–3 greyish-white, spotted eggs.

Adult in flight. Note wings forming a shallow V with silvery flight feathers and black underwings.

Juvenile has greyish skin on head, duller brown plumage and black-tipped bill.

Range *C. a. aura* breeds in western North America to Costa Rica, northern Bahamas and Greater Antilles. North American birds winter through the southern breeding range to central South America.
Status Abundant breeding resident throughout the Cuban archipelago; migrants from North America occur in winter.

Black Vulture

Coragyps atratus 58–68cm

Local name Zopilote.
Taxonomy Monotypic.
Description Entirely black, with grey scaly bare head and legs, very short square tail. In flight rapid flaps with short glides; when soaring, wings are outstretched horizontally with conspicuous white base to six outer primaries.

Voice Usually silent. May emit some grunts.
Similar species Turkey Vulture has flight feathers silvery below, and more soaring flight with wings held in shallow V; tail longer and rounded.
Habitat and behaviour Coastal and open areas. Feeds on carrion. No breeding data for Cuba.

Adult entirely black except for white primaries. Wings held horizontally when flying.

Range Breeds in southern and eastern United States to central Argentina.
Status Very rare visitor, mainly in western and central Cuba. Recently very local and breeding resident population in northern Matanzas province (Lomas de Bibanasí).

Cuban Kite

Chondrohierax wilsonii 38–43cm

Local name Gavilán Caguarero.
Taxonomy Monotypic.
Description Both sexes have cere and facial skin bluish-horn, iris bluish-white; very large, heavy, hooked yellow bill, long tail with blackish and white broad bands, and yellow legs. Male is ashy-grey, head paler, appears whitish from a distance, finely barred below, mostly with grey. Female is dark brown with conspicuous reddish bars on whitish or cream breast and belly; tan on nape. In flight, wings are paddle-shaped with broadly rounded tips and very distinct black barring on underwings; long tail with wide blackish and white bands. Juvenile female has head and upper back speckled with white. See page 371 for an illustration of this species.
Voice Unknown.
Similar species Broad-winged Hawk has smaller bill, pointed wings and shorter tail without wide bands, streaked underparts and unbarred underwings.
Habitat and behaviour Riparian montane forest, also in coastal dry woodland. Breeding season unknown and nest undescribed. Feeds apparently on tree snails.

Range Cuba. Only found in the humid mountains of the Grupo Nipe–Sagua–Baracoa, San Rafael, Yateras, and El Zapote de Mal Nombre, a locality in Guantánamo provinces.
Status Critically Endangered. Endemic species, close to extinction. Only five sightings in the past 30 years in eastern Cuba. Also considered a subspecies of Hook-billed Kite *C. uncinatus*.

Swallow-tailed Kite

Elanoides forficatus 51–66cm

In flight adult shows entirely white head, underparts and underwing; long black pointed wings; and deeply forked tail.

Local name Gavilán Cola de Tijera.
Taxonomy Polytypic (2).
Description Slender silhouette with long pointed wings, bluish-grey upperparts with black band across upper back and inner wing; white head, underparts and underwing-coverts; black primaries and secondaries, black tail very long and deeply forked. Juvenile similar to adult, but duller, with fine streaks on head and breast. Shorter tail.
Voice Loud shrill whistle: *ke-wee-wee* or *je-wee-wee*, the first note very short.
Similar species None.
Habitat and behaviour Wetlands, savannas, river mouths, clearings. Feeds on reptiles (especially canopy-dwelling snakes), frogs, insects, also nestling birds. Usually eats prey while in flight.

Range *E. f. forficatus* breeds from coastal south-eastern United States to Texas; winters in South America and migrates through the northern Bahamas and western Greater Antilles.

Status Uncommon passage migrant, mainly on Cuba and northern cays, July–December, less common in spring, February–April.

Mississippi Kite

Ictinia mississippiensis 37cm

Adult has pale grey head, grey body, dark primaries and long tail.

Local name Gavilán del Mississippi.
Taxonomy Monotypic.
Description Small kite, plain grey overall with paler head, dark area around red eye, bill black, dark grey wings and long square tail. Female with white shafts on outer tail feathers and often white on vent. In flight long pointed wings with short outermost primaries, whitish patch on secondaries from above; underwing-coverts grey with blackish flight feathers on distal half; secondaries with straight trailing edge and white border. Juvenile underparts streaked brown.
Voice Whistled *pe-teeew*.
Similar species None.
Habitat and behaviour Open forests and wetlands. Sweeping flight while hunting. Feeds on flying insects.

Range Southern United States; winters in Bolivia, Paraguay and northern Argentina.
Status Rare passage migrant throughout the Cuban archipelago, September–December and March–April.

Snail Kite

Rostrhamus sociabilis 43–48cm

Local name Gavilán Caracolero.
Taxonomy Polytypic (3).
Description Both sexes have long, slender, hooked, black-tipped bill; red eye, yellow-orange cere and legs. Male is blackish-grey, appearing almost black from a distance. Female is dark brown with forehead, throat and supercilia whitish, and densely streaked breast and belly. In flight shows whitish patch on base of primaries on underwing, broad white band across base of tail and coverts, and broad black subterminal band with pale tip. Juvenile resembles female, but eyes brown, finer streaking on buffy underparts, paler legs and base of bill; back and wings with a scaly appearance caused by rufous feather edges. Tail appears triangular and is often twisted as bird positions itself in the air.
Voice A dry, rasping chatter, *ga-ga-ga-ga*.
Similar species Northern Harrier has narrower wings and white rump (not tail base); flight profile is a shallow V. Cuban Black Hawk has banded tail, heavier and straighter bill.
Habitat and behaviour Freshwater wetlands, reservoirs, rice fields. Flapping flight is slow and leisurely, interrupted by intermittent glides, with broad wings held horizontally, typically low over water. Feeds on apple snails (*Pomacea*), seized with feet from just under the water's surface, eaten on preferred feeding posts. Breeds March–July. Compact nest low in a bush or tree, of twigs, varying from a small flimsy platform to a compact structure; lays 1–3 white eggs, spotted with reddish-brown.

Adult male is entirely blackish-grey in the field.

Adult male in flight. Male and female have white at base of outer primaries, broad white band across base of tail and broad black subterminal band with pale tip.

Juvenile has buffy-tan face, throat and breast, and brown eyes.

Adult female.

Range *R. s. plumbeus* breeds in the Florida Everglades, Cuba and Isle of Pines.
Status Common breeding resident throughout the Cuban archipelago and Isle of Pines.

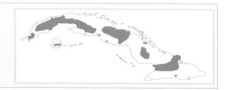

Bald Eagle

Haliaeetus leucocephalus   79–94cm

Local name Águila Calva.
Taxonomy Polytypic (2).
Description Adult with large white head, dark plumage overall and white tail; long, massive yellow bill. Second-year bird is brown with whitish on abdomen, underwing-coverts, flight feathers and tail base. Juvenile has brown underparts, whitish on coverts less marked, white on flight feathers, dark bill.
Voice Weak whistles.
Similar species None.
Habitat and behaviour Along large lakes, lagoons and rivers. Soaring flight on broad wings held evenly horizontal. Feeds on fish. Solitary.

Adult dark overall with white head and neck, whitish tail and large, hooked yellow bill.

Adult in flight.

Range *H. l. leucocephalus* breeds in North America south to north-west Mexico and Florida.
Status Very rare winter visitor in western Cuba in Pinar del Río and Havana provinces, November–March.

Northern Harrier

Circus hudsonius 46–61cm

Local name Gavilán Sabanero.
Taxonomy Polytypic (2).
Description Slender hawk. Adult has owl-like face and small bill, distinctive white rump, long wings and long, narrow barred tail. Male has bluish-grey head, upperparts and upper breast, rest of underparts whitish; in flight shows whitish underwings, black tips to primaries and dark trailing edge to wing. Female larger with dark brown upperparts, wings and tail, and heavy rusty streaking on buffy underparts; in flight underwings are heavily barred. Juvenile similar to female but darker above and cinnamon below.
Voice Silent.
Similar species Gundlach's Hawk has broader wings, a flatter flight profile, fast wingbeats with short glides; rump is unmarked.
Habitat and behaviour Rice fields, marshes, grassy fields. Flies low with tilting flap and glide, hunting for birds, especially young gallinules, stilts and coots. Also in littoral areas, perched on the ground or a low bush hunting for mice, lizards, toads, crabs and insects.

Male has grey upperparts with an owl-like face and a rather long barred tail.

Male in flight.

Female. Note brown upperparts, white rump and streaked underparts.

Range Breeds in North America and northern Mexico; winters south to Panama and the West Indies, rarely to northern South America.
Status Locally common winter visitor and passage migrant throughout the Cuban archipelago, August–April.

Juvenile in flight. Note orange-buff underparts; white rump is only visible from above.

Gundlach's Hawk

Accipiter gundlachi 43–51cm

Adult. Note lightly-coloured barred underparts.

Local name Gavilán Colilargo.
Taxonomy Polytypic (2).
Description Adult has bluish-grey upperparts, underparts whitish finely barred with rufous; iris red. Older birds attain grey underparts with faint blackish streaks and rufous thighs or with just very faint rufous barring on abdomen. Wings long and narrow towards tip; white undertail-coverts may cover rump area; tail long, rounded and banded, from above narrowly tipped with white. Juvenile has white iris, brown upperparts and brown, heavily streaked underparts.
Voice Nest-defence call, a nasal, repeated *kek, kek, kek*; also a soft double whistle during breeding.
Similar species Sharp-shinned Hawk is smaller, with broader, shorter wings and square tail, narrower at base; in flight head does not extend far forward of wings. Female Northern Harrier has white rump; in flight wings are held in a shallow V. Broad-winged Hawk is smaller with shorter tail, shorter broader wings are white below with black trailing edge and black tips to primaries.
Habitat and behaviour Woodland, second growth, pine forest, wetlands, from sea level to mid-elevation. Fast flight with rapid wingbeats and short glides, head extending well beyond wings; usually flies low above the canopy in forest and very high while soaring. Feeds mainly on birds. Breeds February–May, nest a flat twig platform lined with grass at 10m or higher in trees; lays 2–4 pale bluish-green eggs.

Adult female in flight.

Juvenile has brown plumage, pale iris and heavily streaked underparts.

Range Cuba.
Status Endangered globally and nationally. Endemic species with two subspecies: *A. g. gundlachi* is uncommon and local in western and central Cuba and more widely distributed; *A. g. wileyi*, in eastern Cuba and northern cays, is less common and this subspecies is considered doubtfully valid.

Juvenile in flight. Note heavily-streaked underparts, white undertail-coverts, barred wings and banded tail.

Immature in transition to adult plumage.

Juvenile.

Sharp-shinned Hawk

Accipiter striatus 25–35cm

Adult, North American subspecies.

Local name Gaviláncito.
Taxonomy Polytypic (10).
Description Small, dark hawk with small head not extending far in front of wings during flight. Bluish-grey upperparts, whitish underparts with breast and belly finely barred reddish-brown. Female is larger. Light-bodied, with long, narrow, square tail barred with black; short, rounded wings adapted for forest flight. *A. s. fringilloides* has less defined barring than migratory form. Juvenile dark brown above, and may have whitish spots on upperparts and darkly streaked breast and belly. Young birds have paler underwing.
Voice A high, repeated *kik*.
Similar species Gundlach's Hawk is larger and heavier; in flight head extends well forward of wings, tail longer and rounded at tip.
Habitat and behaviour Forests, mostly at moderate to high elevations. Feeds almost entirely on small birds. Like other *Accipiter* species, flies with alternating bursts of shallow, rapid wingbeats and glides. No breeding data for Cuba.

Juvenile, North American subspecies. Note streaked underparts.

Range Breeds throughout the Americas. *A. s. fringilloides* breeds on Cuba and is sedentary. *A.s. velox* breeds in North America and winters south to Panama.
Status Locally threatened subspecies, very rare breeding resident on mainland Cuba. *A.s. velox* is a very rare winter visitor throughout the Cuban archipelago, October–April.

Cuban Black Hawk

Buteogallus gundlachii 51–58cm

Local name Gavilán Batista.
Taxonomy Monotypic.
Description Dark chocolate-brown with pale edges to feathers; cere, base of black-tipped bill and legs yellow to orange. In flight broad rounded wings have white half-moon patch at base of primaries; tail short with broad black and white bands. Juvenile lighter brown, with upperparts spotted, head streaked, white underparts with dark brown streaks and pale finely banded tail.
Voice A repeated whistle suggesting the syllables *ba-tis-ta*, with stress on second syllable.
Similar species Immature Broad-winged Hawk is smaller and less robust with less distinct white markings on face and thinner legs; in flight wings are paler and shorter without distinct white half-moon patch on base of primaries. Juvenile Red-tailed Hawk is larger, with white patch on breast surrounded by distinct black streaks.
Habitat and behaviour Coastal thickets and mangroves. Flies with wings held exactly horizontal when foraging for snails and may show upturned primary tips; also, frequently soars high. Feeds on crabs and birds. Tame, often seen on low perch. Breeds January–July; nest usually 3–6m high; lays three brown-spotted, creamy-olive eggs.

Adult in flight. Note white half-moon patches at base of primaries.

Adult is dark brown with distinctive banded tail.

Adult.

Range Cuba, Isle of Pines and cays.
Status Near Threatened. Endemic species. Widespread and common in coastal areas and cays throughout the Cuban archipelago.

Adult in flight.

Juvenile in flight. Note heavily-barred tail and underwings and heavily-streaked underparts.

Juvenile.

Juvenile.

Broad-winged Hawk

Buteo platypterus 35–41cm

Local name Gavilán Bobo.
Taxonomy Polytypic (6).
Description Large and robust with dark brown upperparts, underparts barred reddish. In flight broad wings are slightly pointed on black tips, underwing white with black trailing edge, tail banded black and white (broad white central band). Juvenile has brown-streaked underparts with duller and more numerous tail-bands.
Voice A two-toned whistle, the second note longer, *kee-deee*.
Similar species Gundlach's Hawk is larger, tail is longer and banded, wings long, rounded and narrower towards tip; white undertail-coverts may cover rump area. Juvenile Red-tailed Hawk is larger, with distinct white patch on breast. Juvenile Cuban Black Hawk is larger, with more distinct white markings on head, thicker legs and, in flight, a white patch on underwing at base of primaries.
Habitat and behaviour Dense, undisturbed semi-deciduous, tropical karstic forest, evergreen woodlands, and pine forests. Soaring flight. Feeds mainly on insects, birds, reptiles. Breeds March–July; nest is a small and loose structure of twigs in trees; lays three white eggs, usually marked with brown.

Cuban subspecies in flight. Note streaked underparts are diagnostic of this subspecies.

Adult (Cuban subspecies) has streaked underparts. Note banded tail.

Adult has dark head, streaked underparts and pale brown eye.

Range *B. p. cubanensis* breeds on Cuba.
B. p. platypterus breeds in central Canada and eastern United States, wintering south from Mexico to Brazil and the West Indies on Cuba, Puerto Rico, and from Antigua south to Grenada and Tobago.
Status *B. p. cubanensis* is a fairly common breeding resident on mainland Cuba. *B. p. platypterus* is frequent in winter and rare on passage on Cuba and Isle of Pines with one record on Cayo Coco, August–May.

Red-tailed Hawk

Buteo jamaicensis   48–64cm

Adult. Note dark brown upperparts and red tail.

Local name Gavilán de Monte.
Taxonomy Polytypic (14).
Description Large hawk, with upperparts dark brown mottled white; underparts vary from whitish-buff to reddish, white area on breast edged with bold dark streaks forming a band across abdomen. In flight adult shows broad wings, underwing has dark bar on leading edge and wrist, and dark primary tips, reddish tail has dark subterminal band; primaries appear well separated when soaring. Juvenile has grey-brownish upperparts, and barred tail.
Voice A raspy *keeer-r-r-r*, slurring downwards.
Similar species In flight Broad-winged Hawk has round wings with black tips to primaries. Adult has black-and-white banded tail.
Habitat and behaviour Semi-deciduous and evergreen woodland, tropical karstic forest, swamp forest, rainforest and pine forest, from sea level to high elevations. Soaring flight often with Turkey Vultures, also flaps frequently.

Adult in flight displays reddish tail and broad wings with dark bar on leading edge and wrist mark.

Juvenile has grey-and-white plumage and pale iris.

Range Breeds from North America to Panama. Mainly sedentary, although northern North American birds migrate into the southern breeding range. *B. j. solitudinis* breeds in Bahamas, Cuba and Isle of Pines.
Status *B. j. solitudinis* is a fairly common breeding resident on Cuba and northern cays.

Crested Caracara

Caracara cheriway 50–63cm

Local name Caraira.
Taxonomy Monotypic.
Description Mostly blackish-brown with bushy crested head, and long neck and legs. Conspicuous black cap and nape; white cheeks and nape; breast and upper back white, finely barred black; facial skin reddish-orange, with heavy bluish bill; legs yellowish-olive. Juvenile pale brown, with whitish throat and some streaks on breast, grey legs. In flight wings wide and rounded, with white patches on outer wing; tail white, finely barred with broad black terminal band.
Voice Alarm call *caracá–caracá*.
Similar species None.
Habitat and behaviour Pastures, palm-covered savannas, swamp forest, mangroves. Feeds on carrion, slow-moving prey such as chicks and turtles; distinctive deep wingbeats on the downstroke and rarely soars, commonly seen on the ground. Breeds March–December; nest is bulky structures of sticks, plant stems and weeds built on top of low palms, large bromeliads and trees; lays 2–3 brown-spotted, chocolate-coloured eggs.

Second-year.

Adult in flight. Note white wingtips and lightly-barred tail with blackish terminal band.

Range Breeds in southern United States to north-western Peru and northern Brazil, Leeward Antilles, Trinidad, Cuba and Islas Marias.
Status Generally rare and local breeding resident on Cuba and Isle of Pines, but fairly common on the northern cays.

American Kestrel

Falco sparverius 23–30cm

Female North American migrant subspecies. Note reddish-brown streaked underparts.

Male North American migrant subspecies. In flight male shows grey wing-coverts and tail with black subterminal band.

Local name Cernícalo.
Taxonomy Polytypic (17).
Description Both sexes have rufous back and tail, two vertical black facial stripes, and false-eye spots on nape. Male has bluish-grey upperwing-coverts, unbarred tail with wide black subterminal band. Female has rufous wing-coverts, barred back and tail. The Cuban subspecies has two colour morphs: white morph has white breast and belly; red morph has rufous breast and belly. Cuban male has unmarked underparts, but may have spots on sides; female frequently has streaks on breast; others may have spotted or barred belly. Juvenile male and female resemble adults but with slightly paler orbital bare skin, cere and legs. Wintering North American birds can be distinguished by their conspicuously streaked or spotted underparts suffused with buff.
Voice A high-pitched, repeated *killy-killy-killy*.
Similar species Merlin has back bluish-grey (male) or dark brown (female), single black facial line, and strongly streaked below; flight usually much faster.
Habitat and behaviour Open country, cultivated lands, often around built-up areas. Often perches on high bare branches, tips of palms and telephone poles. Hovers over open areas with rapid wingbeats, searching for rodents, lizards, large insects such as grasshoppers and dragonflies, and small birds. Wintering birds observed preying on bats. Breeds December–July, builds nest in abandoned woodpecker holes, palms, natural tree cavities; lays 2–5 brown-spotted, cream-coloured eggs.

Female Cuban subspecies, red morph. Note bright rufous underparts.

Male Cuban subspecies, white morph. Note bluish-grey wings and white underparts.

Female Cuban subspecies. Note white underparts.

Range *F. s. sparverioides* breeds in southern Bahamas, Cuba, Isle of Pines and Jamaica. *F. s. sparverius* breeds in North America to central Mexico; winters south to Central America, Panama and Cuba.
Status *F. s. sparverioides* is a common breeding resident throughout the Cuban archipelago.
F. s. sparverius is a common winter visitor and passage migrant mainly on mainland Cuba, August–May.

Merlin

Falco columbarius 25–34cm

Local name Halcón de Palomas.
Taxonomy Polytypic (9).
Description Male dark bluish-grey above. Female brown above with grey uppertail-coverts. Both sexes are buffy below with brown streaks, and poorly defined vertical stripe below eye; dark-barred pointed wings, tail boldly banded light grey and black with white tip. Juvenile similar to female but duller, upperparts with pale-fringed feathers and finer streaking below. Almost invariably appears slightly heavier, darker and faster than American Kestrel.
Voice A whickering trill.
Similar species American Kestrel has rufous tail and back. Peregrine Falcon is larger, with well-defined facial markings.
Habitat and behaviour Open fields, low coastal vegetation, wetlands, pine forest. Fast flight, catching birds on the wing; also hunts rodents and large insects.

Adult female has brown upperparts. Note heavily streaked underparts and faint bar on cheek.

Male has bluish-grey upperparts.

Adult in flight. Note long pointed wings, finely-barred underwings and long banded tail.

Range *F. c. columbarius* breeds in North America, winters to southern United States, Central America and north-west South America and the West Indies. Cosmopolitan.
Status Fairly common winter and passage migrant, sometimes locally common, on Cuba, Isle of Pines, and larger northern cays, July–May.

Peregrine Falcon

Falco peregrinus 38–51 cm

Adult has grey upperparts and wide vertical facial stripe.

Local name Halcón Peregrino.
Taxonomy Polytypic (18).
Description Large and robust falcon. Distinct black vertical facial stripe or 'moustache'. Black head, dark brown to slate-grey above, whitish breast and belly with rufous wash barred with dark brown. Long pointed wings, barred below, long tapered tail faintly barred. Juvenile is brown above, with brown streaks on breast and belly. Males are markedly smaller than females.
Voice A harsh *kak-kak-kak-kak*; also a series of discordant, interrupted notes.
Similar species Merlin is smaller, with less striking facial markings.
Habitat and behaviour All wetlands, cities. Feeds mainly on birds, especially waterbirds; in urban areas mostly on pigeons, occasionally bats. Powerful fast wingbeats; U-shaped dives while chasing prey.

In flight, adult displays long pointed wings, and finely-barred underparts and underwings.

Range *F. p. anatum* breeds in western North America and north-western Mexico; winters in Central America and the West Indies. *F. p. tundrius* breeds in tundra of northern North America and Greenland and winters in South America. Cosmopolitan.
Status *F. p. anatum* is a fairly common winter and passage migrant throughout the Cuban archipelago, with two breeding records. *F. p. tundrius* is an uncommon winter visitor or vagrant, August–May.

Juvenile. Note the streaked underparts.

Northern Bobwhite

Colinus virginianus 23cm

Local name Codorniz.
Taxonomy Polytypic (21).
Description Chunky. Male is beautifully patterned reddish-brown and grey, with broad white supercilium to nape, white throat, blackish collar on upper breast. Female is brown above with whitish-barred abdomen, long buffy supercilium and throat. Both sexes have reddish-brown sides and flanks, and short grey tail. Juvenile similar to adult female but duller.
Voice A bisyllabic whistle, the second part louder and more extended: *bob-WHITE*.
Similar species None.
Habitat and behaviour Grasslands. Gathers in small flocks. Breeds April–September. Nest is a shallow depression lined with grass, with a side entrance, built among grasses; lays up to 18 dull white eggs. Feeds on seeds, fruits, shoots, insects.

Female. Note the brownish-buff throat and wide supercilium.

Male has white throat and broad supercilium that extends to nape.

Range Cuba. Introduced on Hispaniola, Puerto Rico, St Croix, and Andros and New Providence in the Bahamas.
Status Fairly common endemic subspecies, *C. v. cubanensis*, is generally recognised on Cuba, Isle of Pines and Cayo Sabinal, but it is doubtful whether it is a native species.

Black Rail
Laterallus jamaicensis 14cm

Adult. Blackish rail with small white spots on back and wings, and deep-chestnut neck.

Local name Gallinuelita Prieta.
Taxonomy Polytypic (5).
Description Very small; blackish crown, back and wings with small white spots, nape deep chestnut; short, slender black bill; red iris; underparts very dark slate-grey with flanks barred white. Juvenile similar to adults but duller and eye reddish-brown.
Voice Male *kee-kee-keer*; female *whoo-who*.
Similar species Very young rails of other species have entirely black plumage.
Habitat and behaviour Freshwater lagoons and flooded marshes. Strictly nocturnal, solitary and secretive, reluctant to flush. Possibly breeds as there are year-round records. Elsewhere, nest is small, a deep cup of grasses and sedges, with overhanging adjacent grasses or plant stems pulled and woven together to form a concealing canopy. Feeds mostly on insects, crustaceans, seeds.

Range *L. j. jamaicensis* breeds in eastern and central United States and eastern Central America; winters from southern United States to Mexico and Greater Antilles.
Status Near Threatened. Rare winter visitor and passage migrant, July–April.

Clapper Rail

Rallus crepitans 37cm

Local name Gallinuela de Manglar.
Taxonomy Polytypic (19).
Description Heavy-bodied. Blackish-brown upperparts with wide, ashy feather margins; cheeks grey, whitish supercilium and chin, bill long and slightly downcurved, with deep orange on mandible, brownish iris; neck and breast greyish, upper to mid-belly buff, lower belly and flanks faintly barred with white; tail very short, often cocked, undertail-coverts white with variable dark spots; short legs grey with variable orange wash. Juvenile paler below.
Voice Long series of rising and falling *kek* notes.
Similar species King Rail has warmer tones, black on central dorsal feathers edged tawny, rusty wing-coverts, and more distinct barring on abdomen. Virginia Rail is much smaller, with reddish bill and grey cheek contrasting with rusty-brown breast.
Habitat and behaviour Mangrove wetlands. Feeds on crabs, small molluscs, fish, aquatic insects, plants. Breeds April–June, in a bulky cup nest elevated on a firm bank among mangrove roots; lays 5–9 cream-coloured eggs spotted with dark red.

Adult.

Adult. Note greyish breast, upperparts with ashy feather margins, and barred flanks and lower belly.

Range *R. c. caribaeus* breeds in the Greater and Lesser Antilles. *R. c. crepitans* breeds in coastal eastern United States; winters south to Florida and West Indies.
Status Common breeding resident and winter visitor throughout the Cuban archipelago.

King Rail

Rallus elegans 38–48cm

Local name Gallinuela de Agua Dulce.
Taxonomy Polytypic (3).
Description Dorsal feathers have blackish centres with tawny margins; blackish belly and flanks boldly barred white; cheeks, neck and breast rufous, rusty-brown wing-coverts, white chin, red iris. Long bill with orange on mandible; long grey legs; tail very short, frequently cocked. Juvenile is paler below.
Similar species Less richly coloured Clapper Rail has black dorsal feathers with ashy margins and abdomen faintly barred. Virginia Rail (vagrant) is much smaller with grey cheeks.
Voice Repeated *kek-kek-kek, trr-trr-trrrr*.
Habitat and behaviour Freshwater wetlands. Breeds May–December, on the ground in a grass tussock or waterside vegetation; nest is a grassy cup with growing stems pulled over to form a canopy, lays up to nine cream-coloured eggs spotted with reddish-brown. Feeds on seeds, small fruits, insects, crustaceans, frogs, molluscs.

Adult. Note reddish-brown throat, breast and wing. Flank and belly heavily barred black and white.

Cuban subspecies.

Range *R. e. ramsdeni* breeds on Cuba and Isle of Pines. *R. e. elegans* breeds from south-eastern Canada to eastern United States; winters to eastern Mexico (Veracruz) and probably Cuba.
Status Near Threatened. *R. e. ramsdeni* is a common breeding resident on Cuba, Isle of Pines and northern cays; *R. e. elegans* is a rare winter visitor.

Sora

Porzana carolina 14cm

Local name Gallinuela Oscura.
Taxonomy Monotypic.
Description Brown and black upperparts with white streaks, abdomen barred black and white; tail pointed with whitish undertail-coverts. Yellow bill short and stout with greenish tip, short olive-green legs. Breeding adult has black mask, throat and central breast, fading to grey on sides, becoming mainly grey in winter plumage. Juvenile has buffy face and breast; bill duller.
Voice Rising whistled *ker-wee*.
Similar species Yellow-breasted Crake resembles juvenile Sora, but much smaller with black cap and line through eye, bill finer and blackish-olive.
Habitat and behaviour Saltwater and freshwater wetlands. Feeds on grass seeds, small molluscs, worms, insects.

Breeding adult shows black crown stripe, mask and throat.

Adult. Note short yellowish bill and white streaks on brown back.

Range Breeds in North America; winters from southern United States to Peru and West Indies.
Status Common winter visitor and passage migrant on Cuba, Isles of Pines and the northern cays, September–May.

Yellow-breasted Crake

Hapalocrex flaviventer 14cm

Adult. Note very small black eye-stripe, yellowish back streaked white and black-barred lower underparts.

Local name Gallinuelita.
Taxonomy Polytypic (5).
Description Very small; buffy-brown upperparts streaked with black and white; head, breast and neck washed with ochre, black crown and stripe through eye, belly barred blackish and white; short olive bill and yellowish legs.
Voice Repeated thin, high-pitched ascending *peeeeep*.
Similar species Juvenile Sora similar but much larger with plain face and deep-based yellowish bill.
Habitat and behaviour Well-vegetated shores of lakes, lagoons and anthropogenic wetlands; reluctant to flush. Feeds on plants, aquatic and terrestrial invertebrates, and their eggs. Breeds probably April–October. No data for Cuba, but elsewhere nests among aquatic plants, laying 4–5 spotted, pale cream eggs.

Range *H. f. gossii* breeds on Cuba and Jamaica.
Status Very rare breeding resident on Cuba and Isle of Pines.

Zapata Rail

Cyanolimnas cerverai 29cm

Local name Gallinuela de Santo Tomás.
Taxonomy Monotypic.
Description Olive-brown upperparts; white throat, buffy undertail-coverts (not always present) bordered with white; slate-grey underparts with flanks and lower abdomen faintly barred with white in some individuals. Bill olive with red base; red legs. Juvenile duller, lacks red on bill, and legs are olivaceous. See page 371 for an illustration of this species.
Voice Unknown.
Similar species Spotted Rail is black, extensively marked with white.
Habitat and behaviour Dense marshes with low trees and tussock grass, sawgrass and cattail beds. Flight is very weak. Feeds on tadpoles, small vertebrates and aquatic invertebrates. Breeding possibly November–January; nest and eggs not described. (Breeding data, cited in most relevant literature, are ambiguous and do not definitely refer to this species.)

Range Endemic genus confined to the Zapata Peninsula, Cuba.
Status Critically Endangered. Very rare on mainland Cuba. Relict species currently recorded only from Zapata: Santo Tomás (type locality), also sightings from Laguna del Tesoro, Peralta and La Turba.

Spotted Rail

Pardirallus maculatus 28cm

Local name Gallinuela Escribano.
Taxonomy Polytypic (2).
Description Black with white spots on upperparts and breast, wings tawny; belly blackish, barred with white, undertail-coverts white. Bill olive with red base to mandible and bright green base to maxilla; red eye; legs reddish. Juvenile brown overall with duller spots and olivaceous legs.
Voice High-pitched guttural grunt, more like a pig than a bird; also a fast, repeated *tuk-tuk-tuk-tuk*, or short *tuk-tuk*.
Similar species Zapata Rail is olive-brown above, bill slightly shorter with complete red base, white throat and uniform grey below with some faintly barred on lower abdomen.
Habitat and behaviour Freshwater wetlands, sawgrass and rice fields, including in Zapata Peninsula. Feeds on invertebrates, small frogs. Breeds March–December; nest of grass attached to small bush close to water; lays 3–7 spotted eggs.

Adult. Blackish rail with white spots on upperparts and breast; belly barred white; and red legs.

Adult. Note red spot at base of mandible.

Range *P. m. insolitus* breeds from central Mexico to Costa Rica and Cuba.
Status Locally common breeding resident on Cuba, Isle of Pines and northern cays.

Purple Gallinule

Porphyrio martinicus 33cm

Local name Gallareta Azul.
Taxonomy Monotypic.
Description Purple-blue with iridescent green back and wings. Red bill, tipped with yellow and bright pale blue frontal shield; yellow legs. Juvenile entirely buffy-brown, with a green sheen on wings and back; bill mostly dark olive. Both adult and juvenile have white undertail-coverts.
Voice Series of cackling notes interspersed with much deeper *Kr-rruk*, *kek*, *kek*, *kek*.
Similar species Juvenile Common Gallinule is dark grey with white stripes on flanks. Juvenile American Coots are dark grey with white bill.
Habitat and behaviour Freshwater wetlands. Surprisingly adept at swimming, walking over aquatic vegetation, climbing over bushes and flying. Cocks tail frequently while walking. Feeds mostly on seeds, snails, aquatic insects, snails, frogs. Breeds February–September, in a bulky cup nest of dead or green plant material, well above water; lays up to 12 whitish eggs, spotted reddish-brown.

Juvenile has green wash on brown upperparts.

Adult has iridescent bluish-purple head and underparts, pale blue shield and red bill with yellow tip.

Range South-eastern United States and Mexico south to Argentina, and West Indies.
Status Common breeding resident on Cuba, Isle of Pines and Cayo Coco; also a winter visitor.

Common Gallinule

Gallinula galeata   34cm

Local name Gallinuela de Pico Rojo.
Taxonomy Polytypic (7).
Description Slate-grey, brownish on back and wings; bill red with yellow tip; frontal shield red; legs green. Adult has white stripe on sides and undertail-coverts white with black centre. Juvenile is greyish with brownish back, whitish stripe on side; bill and legs dusky.
Voice Clucks, cackles, and harsh cries *kr-r-ru, kruh, kruh, kruh*.
Similar species American Coot is entirely dark grey with white bill. Juvenile Purple Gallinule is buffy-brown, green sheen on back, white undertail-coverts; cocks tail frequently when standing.
Habitat and behaviour Freshwater and brackish wetlands, rivers, rice fields. Feeds on aquatic vegetation, molluscs, worms, fruits. Breeds May–December; nest a bulky platform of dead plant material and debris near or over water or in rice fields; lays up to nine greyish-white eggs, spotted reddish-brown.

Immature is paler than adult. It has grey frontal shield, dull bill with yellow tip and white flank stripes.

Breeding adult has bright red frontal shield, bill and tibial bands.

Range *G. g. cerceris* breeds in the West Indies and Tobago. *G. g. cachinnans* breeds in south and eastern Canada and United States to Panama; Bermuda; Galápagos Islands. Cosmopolitan.
Status Common breeding resident on Cuba, Isle of Pines and northern cays; likely also a winter visitor.

American Coot

Fulica americana 38cm

Local name Gallareta de Pico Blanco.
Taxonomy Polytypic (2).
Description Dark slate-grey; black towards the head.

Adult is grey-black overall and has large feet with lobed toes.

Bill and frontal shield white, except for maroon apex to shield. Legs green, with lobed toes. Both adult and juvenile have white undertail-coverts with black centre. Juvenile paler with white bill.
Voice Typically, *took*, repeated at irregular intervals, and varied clucks and cackles.
Similar species Juvenile Purple Gallinule is buff-brown, with green sheen on back and wings, entirely white undertail-coverts. Juvenile Common Gallinule has thin white stripe on side and brownish bill.
Habitat and behaviour Freshwater and brackish wetlands and reservoirs. May form loose social groups or dense flocks. Submerges for aquatic vegetation, also snails, tadpoles, worms. Breeds May–December; nest a bulky cup of dead leaves and stems of waterside plants near or over water; lays up to 12 eggs, heavily spotted with dark brown and black.

Adult. Note red dot in centre of frontal shield, red eye and partial red ring on white bill.

Range *F. a. americana* breeds in North and Central America; also the West Indies, including Bahamas, Cuba, Jamaica, Hispaniola and Cayman Islands; winters west to Hawaii and south to Colombia.
Status Uncommon breeding resident and common winter and passage migrant throughout the Cuban archipelago.

Limpkin

Aramus guarauna 69cm

Local name Guareao.
Taxonomy Polytypic (4).
Description Large solitary wading bird. Adult has dark brown plumage with white streaks from head to mid-body, long heavy slightly decurved bill, long neck and long dark legs with grey-olive feet. Immature paler and more spotted. In flight appears hunched, the upstroke with a distinct flick and wings barely rising above horizontal.
Voice Highly vocal at dawn and dusk with an unmistakable, unearthly far-carrying *karrao* and a short knocking note.
Similar species Juvenile White Ibis has white underparts and longer bill; immature Glossy Ibis has thinner, more decurved bill and dark unstreaked plumage.
Habitat and behaviour Freshwater wetlands, wet savannas, humid forests. Solitary. Flicks tail constantly. Breeds April–January; large loose nest near water, on the ground or low in trees; 3–6 creamy-buff eggs heavily spotted and blotched with brown, pale grey and lilac. Feeds mainly on apple snails, also invertebrates and frogs.

Adult. Note white streaks on head and neck, which become triangles on body and wings.

In flight Limpkins display long broad wings and fly with head and neck held downwards. Heavy bill is shorter than ibis and only slightly downcurved.

Range *A. g. pictus* breeds in Florida, the Bahamas, Cuba and Jamaica.
Status Common, but low-density, breeding resident on Cuba and Isle of Pines and, rarely, on some northern cays.

Sandhill Crane

Antigone canadensis 86–122cm

Local name Grulla.
Taxonomy Polytypic (5).
Description Large and grey overall, with red featherless patch on forecrown to bill, long straight bill and legs. At rest, the elongated bushy tertials and inner secondary coverts cover the rump area, extending over the tail. In flight neck is extended and wings never rise much below the horizontal; often glides for short distances. Immature has neck and entire head brownish; mottled upperparts.
Voice Rolling *kuroo*.
Similar species Great Blue Heron has yellowish bill and broad black stripes on head; neck is retracted in flight.
Habitat and behaviour Marshes, savannas and grassy fields near pine forests. Breeds February–July; large nest of heaped plant material with central hollow; lays two greyish-green eggs spotted and blotched with reddish-brown.

In flight shows long outstretched neck and darkish tips to primaries.

Adult is grey and has red from forecrown to bill.

Adult pair.

Range *A. c. nesiotes* breeds on Cuba and Isle of Pines.
Status Endangered nationally; endemic subspecies and breeding resident with a restricted range on mainland Cuba; Santo Tomás and San Lázaro in Zapata Peninsula; Jobo Rosado, Santi Spíritus; Ciénaga de las Guayaberas, Ciénaga de Majaguillar in Ciego de Ávila; and Los Indios and La Reforma on Isle of Pines.

Black-bellied Plover (Grey Plover)

Pluvialis squatarola 26–34cm

Local name Pluvial Cabezón.
Taxonomy Polytypic (3).
Description Non-breeding adult and immature have brownish-grey upperparts, prominent black eye, dark ear-coverts, whitish supercilium, whitish underparts lightly streaked on breast, short heavy black bill and dark legs. Juvenile has more pronounced streaking on underparts. Birds in early autumn and late spring show black mottling on face and underparts. In flight shows broad white band across upperwing, diagnostic black axillaries on underwing; white rump, uppertail- and undertail-coverts, tail finely barred. Breeding adult has pale grey crown, white band from forehead and supercilium down sides of neck to shoulder; black face and breast to mid-belly, upperparts mottled black and pale grey.
Voice Plaintive *klee-u-ee* and single *klee*.
Similar species Non-breeding American Golden Plover is smaller and browner with finer bill and longer wing projection at rest. In flight uppertail is dark and black axillaries are absent; breeding adult has black undertail-coverts.
Habitat and behaviour Mainly coastal on sandy and rock shores, brackish and saline lagoons.

Juvenile. Note well-defined white edges to feathers on upperparts, streaked breast and flanks.

Adult in breeding plumage.

Non-breeding adult has brownish-grey crown and upperparts and white abdomen.

Adults in flight. The black axillaries are diagnostic of this species.

Range Breeds in North American Arctic; winters on both coasts from North America to Chile and Argentina, and the West Indies. Cosmopolitan.
Status Common winter visitor and passage migrant throughout the Cuban archipelago, July–May.

American Golden Plover

Pluvialis dominica 26cm

Non-breeding adult. Note dark ear patch, broad white supercilium and light spotting on grey-buff underparts.

Local name Pluvial Dorado.
Taxonomy Monotypic.
Description Non-breeding adult has greyish and gold upperparts, dark crown, whitish supercilium, dark ear-coverts, white underparts with streaking on breast, short black bill, black legs. Breeding shows bright mottled gold and brown upperparts, white S-shaped stripe from forehead to crown and side of breast, entirely black underparts. In flight shows dark rump and tail; unmarked axillaries.
Voice Rich trebled *queedle* in flight.
Similar species Black-bellied Plover is larger, bill heavier; paler ear-coverts, shorter wing projection; in flight shows white rump and black axillaries at base of underwing.
Habitat and behaviour Coastal sandy beaches and saltmarshes, brackish lagoon edges.

Breeding male.

Range Breeds in North America, wintering to southernmost South America, and on passage on North America's Pacific and Atlantic–Gulf coasts, rarely in the West Indies.
Status Rare passage migrant throughout the Cuban archipelago, July–December and January–April.

Snowy Plover
Charadrius nivosus 14–15cm

Local name Frailecillo Blanco.
Taxonomy Polytypic (2).
Description Very small plover with long thin black bill and dark greyish legs. Non-breeding and breeding female have pale sandy-grey upperparts and pale brown breast-side patches. Breeding male has black bar on forehead and ear-coverts, black breast patch on sides. In flight has white sides to black-tipped tail.
Voice A low *krut* and whistled *ca-wee*.
Similar species Semipalmated and Wilson's Plovers are darker above and have complete breast-bands. Piping Plover has shorter, thicker bill and orange legs.
Habitat and behaviour Sandy beaches and coastal saltmarshes. Breeds April–August; lays 2–4 buffy eggs marked with black or pale grey in a sand scrape.

Breeding adult has white collar and forehead, and black forecrown, ear patch and incomplete breast-band.

Non-breeding adult lacks black markings on head. Note thick black bill and sandy-grey upperparts in all plumages.

Range Breeds in western, central and southern United States and Bahamas, Greater and Leeward Antilles. Winters south to Panama.
Status Near Threatened. *C. n. nivosus* is a very rare breeding resident or summer breeding visitor and passage migrant to Cuba and the cays.

Wilson's Plover
Charadrius wilsonia   18–20cm

Local name Titere Playero.
Taxonomy Polytypic (4).
Description Small plover with large head and long, thick black bill; greyish-brown crown and upperparts, white collar and forehead continuous with supercilium to behind eye; white underparts; greyish-pink legs. Non-breeding adult has wide brown breast-band (may appear incomplete). Breeding male has black bar on forecrown, lores and complete breast-band, brown in female. Juvenile has breast-band mottled with buff, upperparts edged pale and dull yellowish legs. In flight adult shows dark outer upperwing with broad white band and black tail.
Voice Sharp *wheet* call.
Similar species Semipalmated Plover has narrow breast-band, very short bill and yellow-orange legs.
Habitat and behaviour Sandy shores and saline lagoon edges, preferring drier areas than Semipalmated Plover. Breeds April to August; in shallow scrape, lays 2–3 creamy eggs spotted or blotched with black and pale grey.

Breeding male has black lores, forecrown and wide breast-band. Note thick black bill in all plumages.

Non-breeding adult and juvenile shows a brown breast-band, lores and forecrown.

Range *C. w. wilsonia* breeds from coastal eastern United States (Virginia) to eastern Mexico, Belize and West Indies. Winters south to coastal northern South America, and in the West Indies throughout the breeding range.
Status A common summer breeding visitor and passage migrant throughout the Cuban archipelago, February–November.

Semipalmated Plover
Charadrius semipalmatus 18.5cm

Local name Titere Semipalmeado.
Taxonomy Monotypic.
Description Non-breeding adult has dark brown crown and upperparts, white forehead, supercilium and collar, dark ear-coverts, narrow dark brown breast-band, very short blackish bill with small amount of orange at base, and yellow-orange legs. Breeding male has black frontal bar and band from eye to ear-coverts, narrow black breast-band extends to form a black collar, bill orange with black tip, legs orange-yellow; breast-band and collar dark brown in female. Juvenile similar to non-breeding with upperparts edged pale, breast-band may be reduced, bill entirely black and legs yellow. In flight shows broad white band along dark outer wing.
Voice High-pitched *koo-wee* or *kweet*.
Similar species Non-breeding Piping and Snowy Plovers are pale sandy-grey above with incomplete breast-band. Wilson's Plover has wider breast-band, heavier longer black bill and greyish-pink legs.
Habitat and behaviour Sandy beaches, brackish lagoon edges and saltmarshes. Forages for worms, grasshoppers, beetles and ants, alternating quick runs with pauses.

Non-breeding adult.

Adult in flight. Note black chest-band, dark brown upperparts and white wing-stripe.

Breeding adult has black neck-band, forecrown and eye patch; bill is orange at base.

Range Breeds in northern North America; winters from United States Pacific and Atlantic–Gulf coasts to southern South America and the West Indies. Non-breeding birds often summer in the wintering range south to South America.
Status Common winter visitor and passage migrant throughout the Cuban archipelago, July–June.

Piping Plover

Charadrius melodus 18–20cm

Non-breeding.

Breeding adult has black forecrown and incomplete breast-band; bill is orange at base.

Local name Titere Silbador.
Taxonomy Monotypic.
Description Small rounded head and very short thick black bill that may have faint yellow-orange at base, white collar and forehead, short orange legs. Non-breeding pale sandy-grey upperparts and pale incomplete breast-band. Breeding has black band on forehead and incomplete black breast-band, bill orange at base and black at tip.
Voice Whistled *peep-lo*.
Similar species Semipalmated and Wilson's Plovers are brown above. Snowy Plover has longer thin black bill and dark greyish legs.
Habitat and behaviour Sandy shores with seaweed cover and saltmarshes. Regularly stands near sea edge.

Adult in flight. Note broad white wing-stripe and white collar; outer wing and tail tip are both dark.

Range Northern and eastern North America, wintering from south-eastern United States to the West Indies in the Bahamas and locally in the Greater Antilles.

Status Near Threatened. Uncommon and local winter visitor and passage migrant on Cuba, most regularly on the northern cays, July–April.

Killdeer

Charadrius vociferus 25cm

Local name Titere Sabanero.
Taxonomy Polytypic (3).
Description Large, long-tailed plover with dark brown head and back, orange eye-ring, white on forehead continues below eye, white supercilium and collar, two black breast-bands on white underparts (one encircles the neck), pointed black bill and pinkish legs. Juvenile is similar but duller with upperparts edged pale. In flight shows orange rump and upper tail, black distally with white tip, and broad white wing-stripe across upperwing.
Voice Call *kil-deer* and *kdee-dee-dee,* alarm *deet* repeated.
Similar species None. Only plover with two breast-bands and orange rump.
Habitat and behaviour Wetlands and, rarely, saline lagoon edges. Breeds March–November; shallow scrape, lays 3–4 eggs marked with black and pale grey. Forages for insects and seeds on grasslands often far from water.

Adult. Kildeer is the only plover to have two breast-bands.

Adult in flight. Note orange rump, wide wing-stripe, and long tail with a black subterminal band.

Range *C. v. ternominatus* breeds in the Greater Antilles, the Bahamas and Virgin Islands, and is sedentary. *C. v. vociferus* breeds from North America to Mexico and winters to northern South America and the West Indies.
Status *C. v. ternominatus* is a common breeding resident. *C. v. vociferus* is a common winter visitor and passage migrant. Both subspecies occur throughout the Cuban archipelago.

American Oystercatcher

Haematopus palliatus 46cm

Adult has black head and neck; very long, thick orange bill; and red eye-ring.

Local name Ostrero.
Taxonomy Polytypic (5).
Description Dark brown upperparts, black head and breast sharply contrasting with white underparts; long sharply pointed bright red-orange bill; legs pale pink. In flight shows white median wing-stripe and uppertail-coverts; black tail. Juvenile has brownish head and neck, dark bill and greyish legs.
Voice Loud whistled *queep*.
Similar species None.
Habitat and behaviour Sandy beaches and brackish lagoons. Breeding April–June, recorded on Cayuelo del Mono, off Cayo Pajonal, Villa Clara province. No nesting data.

Adult. In all plumages, shows white abdomen, tail coverts and underwing-coverts.

Range Breeds on Atlantic coasts of North, Middle and South America south to south-eastern Brazil, and the West Indies.
Status *H. p. palliatus*, previously listed as vagrant, is now a rare breeding resident on the cays and is a rare winter visitor.

Black-necked Stilt

Himantopus mexicanus 34–39cm

Local name Cachiporra.
Taxonomy Monotypic.
Description Large, slim wader with long neck, long flexible, needle-like bill and extremely long salmon-pink legs. Adult is white with black crown and hindneck, white patch over eye; back and wings are black in adult male, brownish-black in female. Juvenile has greyish-brown upperparts edged pale. In flight black upperparts contrast with white underparts, and white with greyish tail, and long trailing pink legs.
Voice Noisy when disturbed, alarm call continuous *yip-yip-yip*, also *kr-eik* when foraging at night.
Similar species None.
Habitat and behaviour Usually in flocks on freshwater and brackish wetlands. Breeds March–August; a scrape with varying amounts of vegetation close to water's edge or outcrops in lagoons; 3–4 buffy eggs with dark markings and spots. Downy long-legged fledglings are fed and defended by highly territorial parents who perform 'broken-wing' distraction behaviour. Takes mainly aquatic insects, crustaceans, worms and water plant seeds.

Juvenile. Note yellow legs, pinkish base to bill and indistinct patch over eye.

Adult female has brownish-black back.

Adult male has black back.

Range Breeds from western and southern United States to Peru and eastern Brazil, Galápagos Islands, and the West Indies from the Bahamas south to Antigua. Winters south through the breeding range.
Status Abundant breeding resident and common winter visitor throughout the Cuban archipelago, October–April.

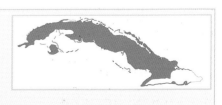

American Avocet

American Avocet 40–51cm

Breeding adult has cinnamon-coloured head and neck.

Local name Avoceta.
Taxonomy Monotypic.
Description Large, slender wader. Non-breeding adult has pearl-grey head and neck, white body, black wings with white band on upperwing-coverts, very long greyish-blue legs, and long black, slender upcurved bill, less pronounced in male. Breeding adult has cinnamon head, neck and breast. Juvenile has brown wash on head. In flight shows long upcurved bill, white head and tail, black-and-white pattern on wings and back, white wing-coverts contrast with black primaries on underwing.
Voice Loud piping *wheeep*.
Similar species None.
Habitat and behaviour Freshwater and brackish wetlands. Breeding noted in June at two small colonies in Ciénaga de Birama, in black mangrove on a beach; nest of herbaceous vegetation, dry twigs, old crab shells and small stones with 2–3 subelliptical to oval eggs, pale brownish-buff with scattered small spots and blotches of black. Feeds on aquatic invertebrates and seeds.

Non-breeding adult has pale grey head and neck. American Avocets are black and white in all plumages.

Breeding adult in flight.

Range Breeds in the Birama Swamp on Cuba, and in south and central Canada, western United States and central Mexico. Migrates south to south and eastern United States, Mexico, the Bahamas and recently on Cuba; vagrant elsewhere in the Greater Antilles.
Status Rare, but increasing, winter visitor and passage migrant on Cuba and Cayo Coco, mainly October–April but occurs in all months. Recently established breeding on mainland Cuba.

Northern Jacana

Jacana spinosa 23cm

Local name Gallito de Río.
Taxonomy Monotypic.
Description Chestnut body, darkening to black towards head, neck and breast; bright yellow frontal shield, bill yellow with white base. Flight feathers are bright yellow-green, wing has a sharp, forward-pointing spur on wrist. Long grey legs with extremely long toes. Female larger and brighter than male, with larger spur. Juvenile is brownish above, white below, thin whitish supercilium and black eye-line, similar yellow-green wing pattern to adults.
Voice Noisy, emits a peculiar loud clacking, usually in flight.
Similar species None.
Habitat and behaviour Freshwater wetlands with dense vegetation. Walks on lily pads and other floating plants. Flight is slow and low, with intermittent bursts of shallow flaps and glides; raises wings over body briefly when alighting. Breeds February–October, in simple nest atop aquatic vegetation. Females often pair with several males, each of which builds a nest, and incubates 2–4 golden-buff eggs heavily scrawled and scribbled with irregular patterns of thin black lines. Feeds on small invertebrates and fish.

Juvenile has long black eye-stripe and white supercilium; bill and frontal shield are pale yellow.

Adult.

Adult. Note greenish-yellow flight feathers.

Range Mexico to western Panama, Cuba, Hispaniola and Jamaica.
Status *J. s. violacea* is a common breeding resident on mainland Cuba, Isle of Pines and some northern cays. Declining.

Greater Yellowlegs
Tringa melanoleuca 33–38cm

Non-breeding adult. Note paler bill at base.

Local name Zarapico Patiamarillo Grande.
Taxonomy Monotypic.
Description Large shorebird with white eye-ring, long bill (1.5 × head length) very slightly upturned in most individuals; long yellow legs. Non-breeding adult has brownish-grey upperparts spotted white, streaked head, neck and breast, bill paler at base, white belly and undertail-coverts, flanks with faint bars. Breeding adult has black and grey upperparts speckled with white, bill dark; head, neck and breast are heavily streaked, flanks are heavily barred. Juvenile upperparts finely spotted with buff. In flight shows dark unmarked upperwing, white rump and uppertail, outer tail barred, and feet extend beyond tail.
Voice Three- or four-syllable *tcheu-tcheu-teu*, descending; alarm call continuous *teu-teu*.

Similar species Lesser Yellowlegs is smaller, straight all-dark bill is same length as head, call usually two syllables.
Habitat and behaviour Wetlands. More active forager than Lesser Yellowlegs.

Range Breeds in south Alaska, central Canada; winters from southern United States to southern South America, and the West Indies. Birds often oversummer in the northern wintering range.
Status Common winter visitor and passage migrant throughout the Cuban archipelago, July–June.

Lesser Yellowlegs

Tringa flavipes 25–28cm

Non-breeding adult. Note bill length is similar to that of the head.

Local name Zarapico Patiamarillo Chico.
Taxonomy Monotypic.
Description Slim, medium-sized shorebird with straight thin all-dark bill (just longer than head) and yellow legs. Very similar to Greater Yellowlegs but smaller and more slender. Non-breeding adult has upperparts brownish-grey with white spots, head may have darker crown, neck and breast finely streaked. Breeding adult has upperparts blackish heavily speckled with white and buff; head, neck and breast heavily streaked, barring on flanks reduced or absent. Juvenile upperparts are brownish with large whitish spots, bill may have yellow base. In flight shows dark unmarked upperwings, white rump and uppertail, and feet extend beyond tail.
Voice One- or two-syllable *tu* or *tu-tu* quieter than Greater Yellowlegs, and alarm *kleet* and continuous *teu-teu*.

Similar species Greater Yellowlegs is larger, bill is longer, pale at base and usually slightly upturned; call three or four syllables and louder. Stilt Sandpiper has greenish legs with bill slightly decurved. Solitary Sandpiper has conspicuous white eye-ring, shorter bill and legs, and dark centre on rump and tail.
Habitat and behaviour Wetlands. Feeding activity is calmer and more vertical.

Range Breeds from Alaska to central Canada; winters from the southern United States to southern South America, and the West Indies.
Status Common winter visitor and common to locally abundant passage migrant throughout the Cuban archipelago, occurs in all months.

Solitary Sandpiper
Tringa solitaria   19–23cm

Local name Zarapico Solitario.
Taxonomy Polytypic (2).
Description Medium-sized shorebird with medium-length fine blackish bill, greenish at base; pronounced white eye-ring continuous with white stripe to eye forming 'spectacles', black lores, olive-green legs. Non-breeding adult has grey-brown upperparts finely spotted white, streaked head, neck and breast, rest of underparts white. Breeding adult is heavily spotted and streaked. Juvenile has brownish wash on head and upperparts. Bounding flight shows dark unmarked upper- and underwings and rump, dark central band on long tail with barred sides.
Voice High-pitched *weet-weet* and *weet-weet-weet*.
Similar species Non-breeding Spotted Sandpiper is smaller, lacks eye-ring and white spots on brown back, short legs greenish or pinkish. Stilt Sandpiper has longer, slightly decurved bill and lacks eye-ring.
Habitat and behaviour Freshwater wetlands, preferring rivers and streams. Solitary, except on migration; walks slowly when foraging. Holds wings up when alighting, folding them slowly; often bobs tail.

Adult in breeding plumage.

Non-breeding adult. In all plumages has white eye-ring, light streaking on head and breast, and upperparts spotted white.

Range *T. s. solitaria* breeds in central Canada; winters from Mexico to Argentina and the West Indies.
Status Common winter visitor and passage migrant throughout the Cuban archipelago, July–May.

Willet

Tringa semipalmata   38–40cm

Local name Zarapico Real.
Taxonomy Polytypic (2).
Description Large heavy-bodied shorebird with pale line above lores; long, thick dark bill paler at base, bluish-grey legs. Non-breeding adult of eastern and western subspecies similar: pale grey upperparts, throat and breast, rest of underparts whitish. Breeding adult of eastern subspecies has shorter and thicker bicoloured bill, pink at base, dark brown upperparts and underparts heavily barred, and buffy flanks. Western subspecies has longer, darker bill and is duller at base. In flight shows broad wings, upperwing black with broad white wing-stripe from secondaries to outer primaries, underwing-coverts black, white rump and grey outer tail. Juvenile has dark crown, grey-brown upperparts; scapulars have black centres.
Voice Around nest *pill-will-willet;* flight and alarm call *yip-yip-yip.*
Similar species None. Black-and-white wing pattern is diagnostic.
Habitat and behaviour Wetlands, beaches, rocky shores. Breeds April–July; on saline lagoon edges in a scrape, lays 2–7 pale green eggs heavily speckled and spotted; adults are very territorial and show distraction behaviour at nest or near fledglings.

Western subspecies in flight.

Non-breeding adult.

Juvenile has brown upperparts and pale wing-coverts.

Breeding adult (eastern race, *T. s. semipalmatus*) has speckled underparts; streaked head, neck and breast; and heavily-barred underparts.

Range Eastern subspecies *T. s. semipalmata* breeds in coastal southern and central Canada to north-east Mexico and West Indies; winters on coast of south and eastern United States south to Brazil. Western subspecies *T. s. inornata* breeds in south and central Canada to western and central United States; winters on coasts of eastern and western United States to northern Chile.
Status *T. s. semipalmata* is a common summer breeding visitor and passage migrant throughout the Cuban archipelago; *T. s. inornata* is a common winter visitor and passage migrant.

Spotted Sandpiper

Actitis macularius 18–20cm

Local name Zarapico Manchado.
Taxonomy Monotypic.
Description Small, short-necked, short-legged shorebird with black eye-line, long white supercilium. Non-breeding adult has brown unmarked upperparts, underparts white with white notch before wing, brown lateral patches on breast, dark bill pale olive at base, legs greenish-olive or yellowish. Breeding adult has white underparts with large black spots (tend to be larger and darker on females), pinkish-orange bill with black tip, upperparts olive-brown with scattered black bars, legs brown. In flight shows brown rump and tail, and short wings with white stripe down mid-wing. Juvenile lacks spots, back and wing-coverts are barred black, legs greenish.
Voice Call *peet* in flight and slow *peet-weet-weet-weet*.
Similar species Solitary Sandpiper has longer neck with more extended posture, dark bill, longer greenish legs and lacks white notch before wing, white spots on back.
Habitat and behaviour Coastal and inland wetlands. Small loose flocks usual on migration, otherwise solitary. Tips hind parts up and down when walking, takes short flights with shallow wingstrokes interspersed with glides and wingbeat below horizontal.

Breeding adult displays heavily spotted underparts.

Juvenile and non-breeding adult has plain underparts and white notch before wing. Juvenile shows barred wings.

Range Breeds in Alaska to Labrador south to central California and North Carolina; winters from southern United States to South America and the West Indies.
Status Common winter visitor and passage migrant throughout the Cuban archipelago, July–May.

Upland Sandpiper
Bartramia longicauda 28–32cm

Local name Ganga.
Taxonomy Monotypic.
Description Large shorebird with small head, short yellow bill with black culmen and tip, large eye with white eye-ring in pale buffy face. Upperparts barred brown with pale edging to feathers; crown and thin neck heavily streaked, breast and flanks buff with dark chevrons; belly unmarked; short wings, long tail and yellow legs. In flight shows unmarked wings with outer wing distinctly blackish, pale inner wing and black rump, underwing-coverts barred black and white.
Voice Rapid liquid *guip-iip-iip*.
Similar species None.
Habitat and behaviour Flooded grassland and agricultural land where it regularly perches on fences; runs and stops when foraging.

Adult. Note small round head, upright stance, and dark chevrons on breast and flanks.

Range Breeds in central Alaska across Canada to north-central and north-eastern United States; winters in eastern Bolivia, southern Brazil to central Argentina. Passage migrant in the West Indies where it is very uncommon to rare in the Bahamas, Cuba, Puerto Rico and the Lesser Antilles.
Status Very rare passage migrant in west and central Cuba, late August–October and March–May.

Whimbrel
Numenius phaeopus 38–46cm

Local name Zarapico Pico Cimitarra Chico.
Taxonomy Polytypic (7).
Description Very large shorebird with long decurved bill (8–10cm) paler at base and striped head. Dark brown upperparts spotted with black and buff, dark crown with pale median stripe, wide pale supercilium, dark eye-line; finely streaked neck and breast, whitish underparts with chevrons on sides, blue-grey legs and feet. In flight appears grey-brown overall with barred underwings on pointed wings. Breeding and juvenile plumages similar.
Voice Rapid whistling *qui-qui-qui-qui-qui*.
Similar species Long-billed Curlew (vagrant) is larger with no head stripes and longer bill; in flight shows bright cinnamon underwing.
Habitat and behaviour Sandy beaches, saline lagoons, saltmarshes and disturbed wetlands, where it probes for crabs on beaches, and picks insects, crustaceans, polychaete worms and molluscs. Usually solitary but small groups occur on migration.

In all plumages, note striped head and chevrons on sides.

Range *N. p. hudsonicus* breeds in north-central Canada; winters on coasts of southern United States to South America and West Indies, where it is uncommon to rare. Cosmopolitan.
Status Rare winter visitor and uncommon passage migrant, most frequent in central mainland Cuba, July–November and February–June.

Ruddy Turnstone

Arenaria interpres 21–23cm

Local name Revuelvepiedras.
Taxonomy Polytypic (2).
Description Stocky, short-legged shorebird with short black, slightly upturned pointed bill. Non-breeding adult has dark brown head with patchy white, dark irregular breast-band, brownish upperparts edged with buff, white underparts and pale orange legs. Breeding male has back and wing-coverts rusty-orange and black, head and neck white patterned with black, black throat and breast-band, orange-red legs; female duller with brownish breast-band. Juvenile similar to non-breeding adult with upperparts edged widely with buff, partial breast-band and dull yellow legs. In flight adult shows rufous, black and white pattern on upperparts and tail, white wing-stripe on upperwing and white underwing with dark trailing edge.
Voice Low rapid guttural *kut-a kut*.
Similar species None.
Habitat and behaviour Sandy beaches, rocky coasts, wetlands. Turns over stones and vegetation probing for prey that includes molluscs and worms; also scavenges in coastal urban areas.

Male in breeding plumage.

Immature.

Non-breeding adult has dark head with white patches and dark upperparts.

Adults in flight. Note broad white wing-stripe and central stripe on back.

Range *A. i. morinella* breeds in Alaska and Arctic Canada and winters from the United States Pacific and Atlantic–Gulf coasts to southern South America and the West Indies. Immature often remain in the wintering range.
Status Common winter visitor and passage migrant throughout the Cuban archipelago, July–June.

SANDPIPERS AND ALLIES 137

Red Knot
Calidris canutus 25–28cm

Local name Zarapico.
Taxonomy Polytypic (6).
Description Medium-sized, heavy round-bodied shorebird, with thick blunt straight bill and short greenish-olive legs. Non-breeding adult has greyish upperparts with scapulars narrowly edged white, whitish supercilium, neck and breast grey-brown with streaks, barring on sides and flanks. Juvenile similar but upperparts scaly. Breeding adult has crown and back grey with streaks, scapulars patterned silver and black with reddish marks; cinnamon-rufous face and throat to upper abdomen. In flight shows diagnostic pale-grey-barred rump and tail.
Voice Low *kuh*.
Similar species Dunlin has longer bill drooping at tip and black legs.
Habitat and behaviour Saline lagoons.

Non-breeding adult is greyish-white with fine streaks on breast; medium-length bill is blunt.

Breeding adult has orange-brown face and underparts.

Range Breeds in the Canadian low Arctic; winters along coast of eastern North America to southern South America. Rare and casual on passage in the West Indies in all months. Cosmopolitan.

Status Uncommon and local winter visitor and passage migrant, most frequently in the northern cays, December–May.

Sanderling

Calidris alba 18–21cm

Breeding adult is reddish-brown with dark-streaked head and breast.

Non-breeding adult is very pale, with black patch on shoulders.

Local name Zarapico Blanco.
Taxonomy Polytypic (2).
Description Small shorebird. Non-breeding adult has the palest plumage of shorebirds with silver-grey upperparts, white face and underparts, black 'shoulder' (lesser coverts), white notch before wing, short black bill and legs; lack of hind toe is diagnostic. Breeding adult has rufous head, chest and upperparts, with dark-streaked head and breast, and bold black mark on back and scapulars, rest of underparts pure white. In flight shows broad white wing-stripe between black forewing and black trailing edge, white sides to tail. Juvenile has dark crown and lores, upperparts blackish and white, breast may be buffy.
Voice Call *wick;* high *ti-ti-ti* among feeding flocks.
Similar species Non-breeding Red Knot is larger, heavier and darker, lacks black shoulder, legs greenish; in flight grey rump and tail are diagnostic.
Habitat and behaviour Sandy, rocky and muddy coasts and saline lagoons. Flocks make rapid synchronised runs along sea edge as they beachcomb for worms, small molluscs and crustaceans.

Non-breeding adult in flight. Note pale, wide, white wing-stripe, black forewing and dark trailing edge.

Range Breeds in north-east Siberia, Alaska and northern Canada; winters to coastal eastern Asia, the Americas and West Indies. Cosmopolitan.

Status *C. a. rubida* is a common winter visitor and passage migrant throughout the Cuban archipelago, August–May.

Semimpalmated Sandpiper

Calidris pusilla   14.0–16.5cm

Local name Zarapico Semipalmeado.
Taxonomy Monotypic.
Description Very small shorebird. Non-breeding adult has greyish-brown upperparts, round head with dark crown and ear-coverts, whitish supercilium, white underparts with lateral streaking on breast, straight black bill broad at base and slightly club-tipped (length overlaps with Western Sandpiper), partly webbed toes and dark olive or black legs. Breeding adult has greyish-brown upperparts with black feather centres and pale edges to scapulars (may show rufous), white underparts with darkly streaked neck and sides of breast, usually not extending to flanks. Juvenile variable, white supercilium, dark ear-coverts, bright greyish-buff maybe with rufous tinge to crown, ear-coverts and scapulars; back and wing feathers have black centres edged pale, giving a scaly effect.
Voice Call loud short *kre-ip*: soft *ti-ti-ti* among feeding flocks.
Similar species Least Sandpiper has finer bill and greenish-yellow legs. Western Sandpiper is paler but similar; bill usually thicker at base, longer, tapering and slightly decurved; legs longer; juvenile and breeding plumages show bright rufous on scapulars.
Habitat and behaviour Freshwater and brackish lagoon edges, beaches, rocky muddy coasts; probes for larvae, worms and molluscs.

Non-breeding adult has dark cap, long supercilium and straight bill.

Adult transitioning to breeding plumage. Note black-and-rufous scapulars and ash-coloured margins.

Non-breeding adult in flight. Note faint wing-stripe and dark central line from rump to tail.

Range Breeds in north-east Siberia, northern North America; winters from Florida, Yucatán Peninsula to both coasts of Central, South America and the West Indies, where it occurs in all months.
Status Common winter visitor and passage migrant throughout the Cuban archipelago, July–May.

Western Sandpiper

Calidris mauri 15–18cm

Local name Zarapico del Oeste.
Taxonomy Monotypic.
Description Almost identical to Semipalmated but slightly larger with bill drooped at tip, most noticeable in females, with thinner tip and thicker base. Non-breeding adult has pale grey upperparts, white underparts with pale finely streaked breast (seldom full breast-band), black bill and legs. Breeding adult has greyish-brown upperparts, bright rufous on crown, scapulars and ear-coverts, white V along the edge of black-rufous scapulars, white underparts heavily streaked on breast with chevrons along sides. Juvenile is greyish with rufous scapulars, dark lores, white underparts with faint streaking on sides of breast.
Voice Usually silent, call *chit*: soft *ti-ti-ti* among feeding flocks.
Similar species Semipalmated Sandpiper has straight bill with club tip, usually shorter but may overlap (difficult to separate); juvenile and breeding plumages have no rufous on scapulars. Least Sandpiper is browner, breast heavily streaked brown, greenish-yellow legs. White-rumped Sandpiper has longer wings extending beyond tip of tail and white rump.
Habitat and behaviour Rocky, muddy shores, beaches, saline lagoons and disturbed mangrove wetlands; when foraging in water often submerges head searching for prey. Often solitary and likely overlooked when in flocks with other 'peeps'.

Non-breeding adult. Note bill which is wider at base and droops slightly at tapering tip.

Juvenile is grey with rufous scapulars.

Range Breeds around the Bering Strait in Alaska (and in Siberia); winters from both coasts of the United States to northern South America and the West Indies, where it is uncommon and occurs in all months.

Status Uncommon to locally common winter visitor and passage migrant throughout the Cuban archipelago, July–April.

Least Sandpiper
Calidris minutilla 12.5–16.5cm

Local name Zarapiquito.
Taxonomy Monotypic.
Description Smallest 'peep'. Non-breeding adult has greyish-brown upperparts and white underparts, entire breast streaked brownish-grey, bill slightly drooped at finely pointed tip, greenish-yellow legs are diagnostic. Breeding adult has dark brown mantle and scapulars with black centres edged rufous or white, dark crown and ear-coverts, darkly streaked head and breast. Juvenile (fresh) is lighter brown with rufous on upperparts, ear-coverts and crown, white supercilium.
Voice Flight call a high trilling *treeep*.
Similar species No other small sandpiper has greenish-yellow legs. Non-breeding Semipalmated Sandpiper has short straight bill, black or dark olive legs and paler markings on breast. Western Sandpiper is greyer above, longer bill drooped at tip, with black legs. White-rumped Sandpiper has longer wings extending beyond tip of tail and white rump.

Non-breeding adult. Note yellow legs that are visible in all plumages.

Habitat and behaviour Wetlands, with freshwater and brackish water preferred, and beaches, rice fields. Forages at water's edge, seldom wading; hunched appearance.

Adult transitioning to breeding plumage. Note white tips to scapulars and wings and back that are edged rufous and white.

Range Breeds in northern North America; winters from southern United States to central South America and the West Indies. Non-breeding birds summer in the wintering range.

Status Common winter visitor and passage migrant throughout the Cuban archipelago, July–May.

White-rumped Sandpiper

Calidris fuscicollis 18–20cm

Local name Zarapico de Rabadilla Blanca.
Taxonomy Monotypic.
Description Small wader, slightly larger than 'peeps', with white rump in all plumages and wingtips projecting beyond tip of tail. Medium-length straight black bill with small red or orange spot at base of mandible is diagnostic but hard to see. Non-breeding adult has greyish-brown upperparts edged whitish, white supercilium, streaked greyish head, breast and flanks, and black legs. Breeding adult has upperparts heavily spotted with grey and black, rufous on crown, ear-coverts and scapulars, heavily streaked breast and flanks. Juvenile has bright rufous and buff on upperparts, crown and ear-coverts, long white supercilium. In flight shows long wings, white rump and dark central tail feathers.
Voice High, thin insect-like *jeet*.
Similar species Semipalmated, Least and Western Sandpipers are all smaller with dark rump and wings that do not extend beyond tail. Least Sandpiper is smaller, browner with greenish legs. Dunlin has longer bill drooped at tip and dark rump.
Habitat and behaviour Mudflats, also freshwater wetlands, rice fields, and flooded grassland; rarely on exposed sandy banks at low tide.

Adult transitioning to breeding plumage. Note rufous tinge on darkly streaked crown, ear-coverts and scapulars.

Adults in flight. Note diagnostic white rump.

Range Breeds in the North American Arctic; winters in southern South America to Cape Horn. Uncommon on passage in the West Indies.

Status Rare passage migrant throughout the Cuban archipelago, July–November and April–June.

Pectoral Sandpiper
Calidris melanotos 20–24cm

Local name Zarapico Moteado.
Taxonomy Monotypic.
Description Heavy-bodied medium-sized shorebird with diagnostic heavy streaks on breast that end abruptly at white belly in all plumages. Non-breeding adult has upperparts brownish and feathers with dark centres and buff edges, white supercilium, black lores, tapered bill yellowish-orange at base and droops at tip, short greenish-yellow legs. Breeding adult has upperparts dark brown edged with rufous and buff; streaking on breast is darker on male. Juvenile similar but brighter, rufous on crown, ear-coverts, scapulars and tertials. In flight wings are practically unmarked with blackish central feathers on rump and tail.
Voice Flocks have noisy flight call, reedy *jrrrrrt*.
Similar species None.
Habitat and behaviour Wetlands, rocky coasts, flooded grassland. Forages mainly in wet mud, seldom in water.

First-winter and juvenile have rufous on crown, scapulars and tertials.

Adult.

Adult. Heavy streaks that end in centre of breast are diagnostic of this species in all plumages.

Range Breeds in the North American Arctic (and north Siberia); winters in Australia and southern South America. Passage migrant through the West Indies.
Status Rare passage migrant on Cuba and northern cays, July–November and January–May.

Dunlin

Calidris alpina 20–23cm

Local name Zarapico Gris.
Taxonomy Polytypic (10).
Description Stocky medium-sized shorebird with relatively long, black bill drooping at tip, short neck and black legs. Non-breeding adult has brownish-grey head and upperparts, lightly streaked breast washed grey; white chin, supercilium and belly, flanks faintly streaked. First-winter bird similar. Breeding adult has rufous on upperparts and crown, large black area on belly, grey-streaked head and breast. In flight distinct white wing-stripe is very broad on primary feathers; rump and tail have brownish-grey central band.
Voice Rough *jeeeep*.
Similar species Red Knot larger with shorter straight bill and greenish-olive legs. Western and Semipalmated Sandpipers are smaller with shorter bills.
Habitat and behaviour Saline lagoons and saltmarshes.

Adult transitioning to breeding plumage. Note black area on abdomen and rufous on scapulars.

Non-breeding adult has greyish upperparts, black legs and long, black drooping bill that narrows towards tip.

Range *C. a. hudsonia* breeds in north-central Canada; winters in coastal eastern United States, Atlantic-Gulf-Caribbean coast to the Yucatán Peninsula, and rarely in the West Indies in the Bahamas and Greater Antilles, Cayman Islands and Lesser Antilles. Cosmopolitan.
Status Rare passage migrant and very rare winter visitor, January–May.

Stilt Sandpiper

Calidris himantopus 20–22cm

Local name Zarapico Patilargo.
Taxonomy Monotypic.
Description Slender, medium-sized shorebird with long black bill thicker at base and drooping at tip, long broad white supercilium and long yellowish-green legs. Non-breeding adult has pale brownish-grey upperparts edged pale; white underparts with faint streaks on neck, breast and sides; flanks unbarred. Breeding adult has rufous crown and ear-coverts, upperparts dark brown marked heavily with rufous and white, neck and upper breast darkly streaked, rest of underparts heavily barred dark brown, including flanks and tail. Juvenile upperparts appear scaly. In flight shows unmarked upperwing, white rump and uppertail, greyish at tip, and feet extending beyond tail.
Similar species Non-breeding dowitchers are larger, with very long straighter bill; in flight white upper rump shows as a wedge between wings, and upperwing shows white secondaries. Both yellowlegs are larger with long yellow legs. Dunlin is stocky with black legs.

Adult breeding. Note heavily-barred underparts.

Voice *Ke-wee* and *ki-wee-wee,* and soft *joof*.
Habitat and behaviour Fresh and brackish lagoons, often foraging in deep water, saltmarshes and saltpans.

Non-breeding adult (left) has white supercilium and fine streaking on underparts. Adult transitioning to breeding plumage (right) has chestnut ear-coverts and bolder streaking on underparts. Note long bill with drooping tip.

Range Breeds in coastal North American Arctic; winters in southern United States to Brazil and northern Argentina. On passage in the West Indies, and locally common in the Greater Antilles in all months.

Status Common passage migrant and uncommon winter visitor on Cuba and the cays, July–December and February.

Short-billed Dowitcher

Limnodromus griseus 26–30cm

Local name Zarapico Becasina.
Taxonomy Polytypic (3).
Description Medium-sized, stocky wader with very long and almost straight, heavy bill, greenish on basal half, blackish at tip, and drooping slightly over outer third of length. Non-breeding adult has greyish-brown upperparts, white supercilium, dark eye-line, fine streaking and spots on breast, pale chin, sides and flanks (although may have densely barred sides), undertail-coverts spotted and barred brown, legs greenish-yellow. Breeding adult has upperparts with black feather centres edged pale and rufous, head streaked with cinnamon wash, neck and breast cinnamon-red with black spots (late spring), sides and flanks heavily barred, belly white. Juvenile has crown, upperparts and tertials with broad rufous-buff edges, and cinnamon-buff wash on breast. In flight shows diagnostic white upper rump as a wedge between wings, pale secondaries, tail barring is variable (may be absent), with white bars as broad as black bars or broader.
Voice Fluid call *tlu-tu-tu*.
Similar species Non-breeding Long-billed Dowitcher is almost identical, but usually bill longer (there is overlap); grey on breast extensive and demarcated. Only accurate separation in the field is by flight call *keek*.
Habitat and behaviour Rocky, muddy coasts and mostly fresh and brackish wetlands. Feeds with up-and-down sewing-machine motion of head while probing for polychaete worms.

Non-breeding adult.

Adult transitioning to breeding plumage. Note reddish-brown underparts, heavy spotting on breast, and barring on sides.

Range *L. g. griseus* breeds in north-eastern Canada; winters from Atlantic coasts of southern United States to Brazil and the West Indies, where common in the Bahamas, Greater Antilles and Cayman Islands; *L. g. hendersoni* is vagrant, probably occurring rarely on Cuba, wintering from south-east United States to Panama.
Status *L. g. griseus* is a common winter visitor and passage migrant throughout the Cuban archipelago, July–May.

Long-billed Dowitcher
Limnodromus scolopaceus   28–32cm

Local name Zarapico Becasina de Pico Largo.
Taxonomy Monotypic.
Description Medium-sized stocky wader, with very long, straight, heavy bill, completely dark or paler at base with greenish tone; white supercilium, dark eye-line. Non-breeding adult has grey-brown upperparts, feathers have dark centres; greyish chin, throat and breast evenly washed with soft grey streaking and almost demarcated; sides and flanks barred buff; undertail-coverts spotted brown; tail barred black and white, with black bars bolder; greenish-yellow legs. Breeding adult has upperparts and scapulars with black feather centres edged white and rufous, head streaked with cinnamon wash, underparts cinnamon-rufous barred with dark brown and white, sides and flanks heavily barred. Juvenile has duller upperparts and tertials edged warm buff. In flight shows white upper rump as wedge between wings, pale secondaries and dark-barred tail.
Voice Flight call light *keek,* singly or in fast series.
Similar species Non-breeding birds are mostly indistinguishable from non-breeding Short-billed Dowitcher which is smaller, slimmer, with shorter legs, bill usually shorter (but there is overlap), fine breast streaking, with spotting on breast-side, more spotting than barring on flanks and tail; only safely separated by flight call.
Habitat and behaviour Wetlands, mostly freshwater and anthropogenic areas.

Non-breeding adult has darker centres to back feathers compared with Short-billed Dowitchers, and even grey wash over breast.

Range Cosmopolitan. Breeds in western North American Arctic; winters from the southern United States to Guatemala and, rarely, in the West Indies.

Status Rare irregular winter visitor and passage migrant on Cuba and northern cays, November–April.

Wilson's Snipe

Gallinago delicata 27–29cm

Adult.

Local name Becasina.
Taxonomy Monotypic.
Description Stocky, short-legged wader with exceptionally long, straight bill and yellow legs. Adult has dark brown upperparts, long whitish or buffy-gold stripes on crown and face continue over back, buffy breast finely speckled, sides and flanks heavily barred, white belly, spotted undertail-coverts. Juvenile similar. In flight shows short orange-tipped tail and short pointed wings with dark underwing.
Voice Alarm call harsh *kcaaap* when flushed.
Similar species None.
Habitat and behaviour Freshwater and anthropogenic wetlands, flooded grasslands. Freezes or explodes from cover into a zigzagging flight on rattling wings.

Adult. Note long white stripes on back, cryptic plumage and very long, straight bill.

Range Breeds in northern and western North America; winters in United States to north-west South America and the West Indies in the Greater Antilles and Cayman Islands. Cosmopolitan.
Status Common winter visitor and passage migrant on Cuba, Isle of Pines and northern cays, August–May.

Pomarine Jaeger (Pomarine Skua)

Stercorarius pomarinus　53cm

Local name Estercorario Pomarinus.
Taxonomy Monotypic.
Description Mostly very dark brown with broad wings, and hooked bill pale at base with dark tip. Two central tail feathers elongated, twisted and spoon-shaped. Light-morph birds have black cap with yellow wash on nape and sides of neck, whitish below with dark band across breast and dark barring along sides and flanks. Dark-morph birds are dark brown below. In both morphs, there is a conspicuous white patch at base of primaries, more extensive below than above; additional pale patch at base of primary coverts on underwing is diagnostic. Bill thick and bicoloured. Non-breeding plumage has barring on rump and undertail-coverts, narrow light barring on back, short, tail-streamers. Juvenile has square-ended short central tail feathers; light morph is neatly barred below.
Voice Usually silent in winter.
Similar species Parasitic Jaeger is smaller, with pointed and rather short tail-streamers; wings appear narrower at base, with less conspicuous single white patch at base of primaries.
Habitat and behaviour At sea off north coast. Flight gull-like, somewhat more leisurely than Parasitic Jaeger; engages in spectacular twisting and diving pursuit of other seabirds. Feeds on fish, offal, carrion.

In flight displays wings that are significantly broader at base, and central tail feathers that are broad and blunt.

Range Breeds in tundra of northern Russia, Siberia, Alaska and Canada, and winters at sea close to coasts between Tropic of Cancer and the equator, and around Australia.
Status Very rare winter visitor in seas off northern coast of Cuba and northern cays, October–April.

Parasitic Jaeger (Arctic Skua)

Stercorarius parasiticus　43cm

Local name Estercorario Parasitico.
Taxonomy Monotypic.
Description Central tail feathers sharply pointed. Light-morph birds have dark cap, whitish at base of forehead, bill black with pale base; white breast (pale breast-band may be present) and upper belly, and grey lower belly. In flight upperwing shows several white shafts, as Pomarine Jaeger, but less conspicuous; underwing has only one defined white wing-patch. Dark-morph birds entirely dark brown below. Non-breeding with barred rump and undertail-coverts, narrow light barring on back; light morph has more extensive and uneven markings on underparts; black cap is less sharply defined and central tail feathers are shorter. Juvenile barred dusky below, more heavily on undertail-coverts; central tail feathers protrude visibly.
Voice Usually silent in winter.
Similar species Pomarine Jaeger is larger, with significantly broader wings; projecting central tail feathers broad and blunt and breast-band is broad and dark. Pale area at base of underwing primary coverts is diagnostic.
Habitat and behaviour Coastal and at sea. Flight is fast, falcon-like. Feeds mostly on fish, largely by piracy.

Adult. Wings appear narrower at base.

Range Circumpolar oceanic breeder in coastal tundra, mainly at 57–80°N, and winters in Southern Ocean, close to coasts of South America, South Africa, southern Australia and New Zealand.
Status Very rare passage migrant to coasts of Cuba and northern cays, July–December and March–April.

Laughing Gull

Leucophaeus atricilla 38–43cm

Breeding adult has entirely black hood, split white eye-ring and dark red bill.

First-winter has greyish-brown head and breast, and brown upperwing-coverts.

Local name Galleguito.
Taxonomy Polytypic (2).
Description Breeding adult has black hood to nape, white crescents above and below eye; dark red bill has curved maxilla that appears slightly downcurved; dark or reddish legs (high breeding). Non-breeding has hindcrown streaked greyish-brown; white face, underparts and tail; dark grey mantle with black outer primaries; and black bill. First-winter is greyish-brown on crown, nape and sides of breast; grey mantle; long wings with brownish coverts; wide dark tail band and white rump. Second-winter similar to non-breeding adult, tail almost white. Adult plumage attained in third year.
Voice Short *che-waa,* and long loud *ha-ha-ha-ha* series from flocks.
Similar species Winter Bonaparte's Gull is small and tern-like, fine straight bill, black spot on ear-coverts and pinkish legs.
Habitat and behaviour Sea coasts and wetlands. Breeds April–July in mixed colonies with terns; nest of feathers, seaweed and debris on sand or rock; lays 2–4 olive-brown-spotted eggs.

Breeding adult in flight.

Non-breeding adult in flight has white tips to black primaries, entirely white tail and white trailing edge to wing.

Range *L. a. atricilla* breeds in the West Indies in large colonies in the Bahamas, Cuba, Virgin Islands, Puerto Rico and Anguilla, and smaller colonies in Leeward Antilles, islands off Venezuela and French Guiana.
Status Common breeding resident, winter visitor and passage migrant throughout the Cuban archipelago.

Bonaparte's Gull

Chroicocephalus philadelphia 33cm

Local name Galleguito Chico.
Taxonomy Monotypic.
Description Tern-like small gull, with rounded head, long wings and thin black bill. Non-breeding has mantle pale grey, white wedge on outer wing and mostly white outer primaries edged with black; blackish ear-spot and two bands over crown; legs pink. Breeding has head with slate-black hood and orange-red legs. Juvenile is brown above with dark carpal bar. In flight upperwing of first-year has dark brown carpal bar, trailing edge and outer primaries, and black subterminal tail-band.
Voice Harsh, somewhat nasal *tiar*.
Similar species Laughing Gull non-breeding has dark grey mantle, with black wingtips and white trailing edge. Black-headed Gull (vagrant) is larger with red bill and dark underwing.
Habitat and behaviour Coasts and inland freshwater ponds. Rapid wingbeats. Feeds on fish, crustaceans, worms.

First-year bird has black ear-spot, dark carpal bar, leading outer primaries and trailing edge to wings, and white tail with black subterminal band.

Non-breeding adult in flight. Note dark spot on ear-coverts, white leading edge to wing and black-tipped outer primaries.

Range Breeds from south-west Alaska to south-east Canada; winters along Pacific coast south to Mexico, on Atlantic coast to Florida and, irregularly, the Greater Antilles.
Status Very rare and local winter resident on the Cuban archipelago, August–March.

Ring-billed Gull

Larus delawarensis 46–51cm

Local name Gallego Real.
Taxonomy Monotypic.
Description Non-breeding adult has very fine brown streaks on crown and nape, pale grey mantle, slender yellow bill with black subterminal ring and yellowish legs; white tail. In flight pale primaries are black-tipped with small white 'windows' on outer tip. Breeding adult has pure white head, bright yellow iris with red orbital ring, red gape. First-winter bird has speckling on head, neck, back and breast, grey mantle, brownish wing-coverts, dark eye, bill flesh-coloured with black tip, legs pink-coloured; in flight dark outer primaries are unspotted, tail has dark terminal band. Second-winter bird like adult with broad, complete ring on bill, some speckling on crown and nape, remnant black band on tail. Adult plumage attained in third year.
Voice Silent.
Similar species Herring Gull larger with pink legs, pink spot on mandible and ring at tip of bill is never complete.
Habitat and behaviour Sea coast bays, saline and brackish lagoons, reservoirs. Usually solitary.

Breeding adult. Note yellow bill with black ring.

Non-breeding adult has very fine brown streaks on crown and nape.

First-winter bird has pinkish bill and legs, black-tipped bill, brownish wing-coverts, dark eye, and no white spots on primaries.

Second-winter bird has grey coverts and pale eye; bill and legs greenish-yellow.

Range Breeds in southern Canada and northern United States; winters from southern United States to Central America and the West Indies in the Greater Antilles and, increasingly, in the Lesser Antilles.

Status Uncommon but regular winter visitor and passage migrant on Cuba and northern cays, October–May.

Herring Gull

Larus argentatus 56–66cm

Local name Gallego.
Taxonomy Polytypic (5).
Description Non-breeding adult has fine streaking on head and neck, pale grey mantle and coverts, heavy yellow bill with pink spot on mandible and may have some black on bill; pink legs in all adult stages. In flight black tips to primaries with white 'windows', white tail. Breeding adult has pure white head and neck, pale iris with orange-red orbital ring and yellow bill with red spot on mandible. First winter bird is entirely speckled and streaked dark greyish-brown, paler on head, bill blackish or blackish-pink with black tip, brown iris; in flight brownish overall with dark brown flight feathers, contrasting pale 'window' on inner primaries, and dark brown tail with mottled brown rump. Second-winter is brownish overall with pale grey inner primaries on upperwing, greyish-white rump and broad black terminal band on tail, bill pinkish with black tip, eye brown. Third-winter has dark grey back and wings, tail mostly white, and blackish subterminal band on yellow bill. Adult plumage attained in fourth year.
Voice *Kyow kyoh kyoh, kya kya kya.*
Similar species Non-breeding Ring-billed Gull is smaller, bill with black subterminal band and yellow legs.
Habitat and behaviour Beaches and rocky coasts.

Non-breeding adult has streaks on head and neck, red spot on bill and pink legs.

Second-winter bird has pinkish-grey bill, black subterminal band and mottled brown mantle and body.

Non-breeding adult has fine streaking on head and neck. in flight adults show pale grey mantle and black primaries with white 'mirror' on outer primaries.

Range *L. a. smithsonianus* breeds from central Alaska to Newfoundland and Great Lakes; winters south to Central America and the West Indies, uncommon in the Bahamas, Cuba and Cayman Islands, rare elsewhere.

Status Uncommon winter visitor and passage migrant on Cuba, Isle of Pines and northern cays, September–May; the majority are first- and second-year birds.

Lesser Black-backed Gull
Larus fuscus 53cm

Second-winter has dark bill, streaks on head and pinkish-yellow legs.

Breeding adult in flight. Note black outer primaries with a distinct white marking on outermost feather.

Local name Gallego de Espalda Negra.
Taxonomy Polytypic (5).
Description Slender long-winged gull. Non-breeding adult has brown streaks on white head and neck, dark grey mantle, thick yellow bill with red spot on mandible, pale eye and yellow legs. In flight shows white on leading and trailing edges of wing, black outer primaries and white tail. Breeding adult has head, neck, underparts and tail entirely white. In flight shows white trailing edge and black outer primaries with distinct white 'window' on outermost feather. First-winter bird is patterned brown and white with streaking on head, neck and underparts, blackish bill, brown wings with two distinct darker bars, whitish rump, white tail with wide black terminal tail-band and pinkish-yellow legs. Second-winter has greyish mantle, grey-and-brown wing-coverts, light streaking on head, breast and sides, dark bill paler at base, dark eye, black tail-band and white rump. Third-winter bird has grey mantle, streaking on head and neck only, rest of underparts white, central blackish band on yellowish bill, yellowish eye.
Voice A strident *kyow*.
Similar species American Herring Gull and Ring-billed Gull both have pale grey mantle and wings; Herring Gull adult has pink legs. Great Black-backed Gull is much heavier and larger with huge bill.
Habitat and behaviour Beaches and coasts.

Third-winter bird in flight. Note speckling on head and neck; bill pale yellow with black central band and legs pinkish-yellow.

Juvenile in flight is brown overall with dark bill and eye, white rump and dark band on tail.

Range *L. f. graellsii* breeds in Iceland, Faroe Islands, western Europe; winters in West Africa and on Atlantic coast of North America.

Status Rare and local winter visitor on Cuba and northern cays, November–April.

Great Black-backed Gull
Larus marinus   76cm

Local name Gallegón.
Taxonomy Monotypic.
Description: World's largest gull, thickset body, white with blackish mantle; huge, powerful bill with red spot on angle of mandible, yellow in breeding plumage, duller in winter; eye yellow, short pinkish legs. In flight broad blackish wings have two white mirror spots on outer primaries, and white trailing edge. Juvenile has mantle checkered in white and blackish-brown, black terminal tail-band, white rump, lightly streaked underparts and barred brown undertail-coverts; bill dark with pale tip. First-winter birds have whiter head, dark bill. Second-winter birds have bill paler at base and browner secondary coverts. Third-winter bird has faint streaking on head and breast, dark and red subterminal marking on bill, dark on tail reduced, back greyish, wings dark brown.
Voice A hollow *cowwp*.
Similar species Adult Herring Gull is smaller with pale grey mantle; juvenile and first-winter are darker overall.
Habitat and behaviour Inshore waters and sea coasts. Feeds on fish, squid, offal.

Breeding adult. Note pale eye with red orbital ring, and heavy, thick yellow bill with red spot.

Range Breeds in the Great Lakes, eastern Canada and United States south to North Carolina; winters in eastern North America south to the Gulf of Mexico and rarely to the West Indies. Northern Hemisphere.

Status Very rare winter visitor to sea coasts of Cuba and northern cays, December–March.

Gull-billed Tern
Gelochelidon nilotica 33–38cm

Non-breeding adult. Note small blackish mask behind eye.

Local name Gaviota de Pico Corto.
Taxonomy Polytypic (6).
Description Heavy-bodied tern with thick short, gull-like black bill, blackish legs, silvery-grey mantle, and shallow-notched tail. Non-breeding adult has fine speckling on head with small blackish mask behind eye; in flight appears white with blackish trailing edge to long curving outer primaries. Breeding adult has black crown to eye and nape. First-winter bird has brownish streaks on mantle and scapulars; in flight outer primaries are darker.
Voice Usually silent; alarm *kee-wak*.
Similar species All other medium-sized black-capped terns have longer, thinner bill and more forked tail. Forster's Tern has pronounced black mask over eye.
Habitat and behaviour Usually solitary on marine sounds and brackish lagoons. Flies like a gull with slow wingbeats, picks prey (invertebrates and small lizards) from water surface or land; rarely plunge-dives. Breeds May–July; 1–2 eggs on ground.

Breeding adult has black cap. In all plumages, species has black bill and legs.

Range *G. n. aranea* breeds in eastern coastal United States, sparsely to Mexican Gulf coast, and in the West Indies in small colonies in the Bahamas, Turks and Caicos Islands, Puerto Rico and likely Cuba (but no data); winters from Central America to Brazil and West Indies.
Status Uncommon breeding visitor with two recent small colonies on Cayo Coco and Cayo Sabinal; uncommon, occasionally locally common, winter visitor and passage migrant on Cuba, Isle of Pines, and northern cays, year-round.

Caspian Tern

Hydroprogne caspia 48–58cm

Juvenile has streaked cap and brownish-grey V-shaped pattern on mantle and wing-coverts.

Breeding adult. Note black cap, red bill, and pale grey mantle and wings.

Local name Gaviota Real Grande.
Taxonomy Monotypic.
Description Largest tern, thickset; heavy deep bill, pale grey mantle, short white tail forked slightly and black legs. In flight shows diagnostic dark outer third of underwing and dark trailing edge to outer primaries on upperwing. Non-breeding adult has crown streaked black and white, darker behind eye; red bill tapers to blackish with yellow tip. Breeding adult has black crown and nape and entirely dark red bill. First-winter bird has crown and bill like winter adult but with brownish-grey scaling on mantle and coverts.
Voice A harsh *kraaow*.
Similar species Non-breeding Royal Tern has white forecrown and black cap on hindcrown. In flight underwing shows outer primaries with dark trailing edge, darker primaries on upperwing and deeply forked tail.
Habitat and behaviour Solitary and in flocks on saline lagoons, inshore waters, mudflats and beaches. Strong, shallow wingbeats on broad wings; hovers and plunge-dives for fish.

First-winter bird has streaked crown, yellow-orange bill, pale legs and tail feathers, darker at tip.

Range Breeds in North America and locally on the Atlantic–Gulf coast and western Mexico; winters mainly in southern United States breeding range and Mexico, rarely to southern Central America and northern Colombia, and in the West Indies in the Bahamas and western Greater Antilles (common on Cuba); very rare to vagrant elsewhere. Cosmopolitan.
Status Common winter visitor and passage migrant on Cuba and northern and southern cays; occurs mainly October–June, but likely in all months.

Royal Tern

Thalasseus maximus 46–53cm

Non-breeding adult displays black cap.

Breeding adult has black crested cap.

In flight first-winter bird displays dark trailing edge to secondaries, dark primaries on upperwing and dark on outer tail.

Local name Gaviota Real.
Taxonomy Polytypic (2).
Description Large crested tern. Non-breeding adult (observed all year) has white forehead and crown with black cap from behind eye, pearl-grey mantle, white underparts and tail with moderate fork, large bright orange bill and black legs; in flight dark trailing edge to outer primaries on underwing and white tail. Breeding adult (April–July) has black crown to below eye and hindcrest and orange-red bill. Juvenile and first-winter birds have head similar to non-breeding adult with yellow bill and legs; juvenile has brown streaking on mantle and scapulars and dark carpal bar. In flight first-winter's upperwing shows dark trailing edge to secondaries, dark primaries and tail with black outer edges.
Voice Call high *kri-ii-iip*.
Similar species Caspian Tern has streaked forehead, heavier red bill, tail less forked and dark outer primaries on underwing.
Habitat and behaviour Usually in groups on estuaries, beaches, saltmarshes and rice fields in coastal areas; also in harbours. Fast, direct flight on long wings; feeds by plunge-diving and dipping in shallows. Breeds March–July; a sandy depression in mixed larid colonies; lays 1–3 creamy-white eggs speckled black and grey.

Range *T. m. maximus* breeds in North America on Pacific and Atlantic–Gulf coasts to Atlantic coast of South America, and in the West Indies, where a small population (*c.* 500 pairs) breeds in the Bahamas, Hispaniola, Cuba, Turks and Caicos Islands, Anguilla, Virgin Islands and Puerto Rico, and where it is also a common non-breeding resident. Winters south through the breeding range to Peru and Argentina, also to West Africa.
Status Common to locally common breeding resident mainly on the northern cays; winter visitor

and passage migrant throughout the Cuban archipelago; immatures and non-breeding adults present in all months.

Sandwich Tern

Thalasseus sandvicensis 41–46cm

Local name Gaviota de Sandwich.
Taxonomy Polytypic (3).
Description Medium-sized crested tern, appears white in flight, with long slender black bill and yellow tip in most plumages, pale grey mantle and black legs. Non-breeding adult has white forehead and forecrown, shaggy black hindcrown; in flight long, pointed wings are held at sharp angle, with dark trailing edge to outer primaries on upperwing, and white tail moderately forked. Breeding adult (May–July) has crested black crown and pinkish wash on underparts. Juvenile has brown scribblings on mantle and scapulars; bill may be entirely dark. First-winter bird has bill dark or with faint yellow tip, upperwing shows blackish secondaries, dark primaries and coverts, dark sides to tail.
Voice Grating *skree-ik* and high-pitched agitated series.
Similar species Gull-billed Tern has shorter, thick black bill and less deeply forked tail. Royal Tern has entirely orange or yellow bill.
Habitat and behaviour Reefs, sea coasts and bays, sandy beaches. Hovers and plunge-dives from considerable height for flying fish. Breeds March–July, on Cayo Los Ballenatos and Cayo Felipe de Barlovento in northern cays with other terns and Laughing Gulls; lays 1–3 creamy-white eggs with dark markings in sandy depression.

Breeding adult in flight displays black cap, white forked tail and black on trailing edge to primaries.

Non-breeding adult has white forehead and forecrown, and black cap. All plumages have yellow tip to black bill.

Breeding adult has long crown forming a crest.

Range *T. s. acuflavidus* breeds on United States Atlantic–Gulf coast, cays off the Yucatán and Belize and the West Indies, mainly in the Bahamas and Greater Antilles; winters in the West Indies and south to Peru and Uruguay. Cosmopolitan.
Status Common summer breeding visitor on cays, and passage migrant throughout the Cuban archipelago, February–September; very uncommon winter visitor, with majority juvenile and first-winter birds.

Roseate Tern

Sterna dougallii   35–41cm

In flight adult displays black cap, pale grey back and wings, long thin bill and forked tail.

Local name Gaviota Rosada.
Taxonomy Polytypic (5).
Description Very long, slender, pure white tern with rounded head, pale grey mantle and long thin bill. Non-breeding adult has white forehead and forecrown, black hindcrown and bill, red legs; in flight upperwing mostly white with dark outer primaries, underwing white with dark-tipped trailing edge on outer primaries, and deeply-forked white tail with long streamers (extending beyond folded wings at rest). Breeding adult has black crown and nape, underparts may be washed pink, bill has basal two-thirds red and black tip briefly in peak breeding season (often just red at base). Juvenile has dark greyish crown, brown scallops on back, dark carpal bar, black bill and legs, shorter tail.
Voice Usually silent; call *chi-vik*.
Similar species Common Tern has less forked tail with dark sides and shorter tail, black legs; juvenile has no scalloping on back. Non-breeding Forster's Tern has oval black eye-patch on white head, and when breeding shows silvery-white primaries on upperwing.
Habitat and behaviour Saline lagoons, marine sounds, inshore waters. Fast shallow wingbeats; makes plunge-dives for fish from high up. Breeds March–July in small numbers in mixed tern colonies with Least and Common Terns on a few northern cays; lays 1–2 creamy eggs, spotted blackish-brown, on bare rock.

Breeding adult has black crown and nape, thin red bill that is black towards tip, and red legs. Tail extends beyond wings.

Range *S. d. dougallii* breeds from Nova Scotia south along Atlantic–Gulf coast to Honduras, cays off Venezuela and the West Indies. Winters at sea in mid-Atlantic and off western Africa.
Status Rare summer breeding visitor on northern cays, March–October.

Common Tern
Sterna hirundo 33–40cm

Local name Gaviota Común.
Taxonomy Polytypic (4).
Description Pale grey mantle, white underparts, black legs; white tail deeply forked with blackish outer rectrices. In flight shows broad blackish trailing edge on outer wing; above, dark area can extend to the base of outer primaries. Non-breeding adult has white forehead, black hindcrown and nape continuous with black area behind eye. Breeding adult has black crown and nape, grey wash to underparts, red bill with black tip, red legs. Juvenile, first-winter and non-breeding plumages have diagnostic black carpal bar and white underparts, black bill; juvenile (and some first-winters) have red legs.
Voice Harsh *kee-aarr*.
Similar species Roseate Tern is whiter, tail deeply forked. Forster's Tern non-breeding plumage has oval black eye-patch; breeding plumage shows silvery-white primaries on upperwing.
Habitat and behaviour Coastal, saline lagoons, rice fields, saltmarshes. Plunge-dives for fish and follows fishing boats. Breeds April–July on a few northern cays, but few data.

Breeding adult in flight. Note black cap, red bill with black tip, grey breast and belly, dark trailing edge to outer primaries, and white tail with dark outermost feathers.

First-winter bird has black carpal bar, black bill and red legs.

Range *S. h. hirundo* breeds in North America and locally to northern South America, and in the West Indies where the total population is small; winters in South America and West Indies, where subadults remain for several years. Cosmopolitan.
Status Rare summer breeding visitor on Cayo Mono, and uncommon passage migrant in all months on the Cuban archipelago.

Forster's Tern

Sterna forsteri 35–42cm

Non-breeding adult in flight. Note black mask over eye. In all plumages, species has dark trailing edge to outer primaries that is visible on underwing. Upperwing shows white primaries with faint black edge.

Local name Gaviota de Forster.
Taxonomy Monotypic.
Description Pale grey mantle, white underparts, reddish legs, deeply forked grey tail with white outer rectrices. Non-breeding adult has white head with diagnostic oval black mask around and behind eye, black bill; first-winter birds have speckled crown. In flight shows silvery-white primaries on upperwing (may be tipped black) and black on outer primaries on underwing. Breeding adult has black cap from forehead to nape, thick orange-red bill with black tip and orange legs. Juvenile has brownish on crown and back, reddish-black bill.
Voice Silent.
Similar species No other non-breeding tern has black ear-coverts patch. Gull-billed Tern has small black mask behind eye. Tail of Roseate Tern is more deeply forked. Common Tern non-breeding and juvenile plumages have black carpal bar.
Habitat and behaviour Saline and freshwater lagoons, saltmarshes.

Non-breeding adult. White crown and black mask are diagnostic of this species in winter.

Range Breeds in the interior and along both coasts of North America and Atlantic–Gulf coast to north-eastern Mexico; winters on Pacific and Atlantic–Gulf coasts to Panama and rarely to the West Indies, in the Bahamas, Greater Antilles and Cayman Islands.
Status Rare to uncommon winter visitor on the Cuban archipelago, August–April.

Least Tern

Sternula antillarum 22–24cm

Breeding adult in flight. Note two black outer primaries.

Breeding adult has black-tipped bill and white forehead.

Local name Gaviotica.
Taxonomy Polytypic (3).
Description Smallest white tern with pale grey mantle, white underparts. Non-breeding adult has forecrown streaked grey, black hindcrown, nape and ear-coverts, black bill (in transition, bill is blackish at base), greyish-yellow legs, long narrow wings, short tail slightly forked. Breeding adult has white forehead, black crown and nape joins black eye-line, long yellow bill slightly drooped at black tip, yellow legs; in flight shows black wedge on leading edge of wing. Juvenile has upperparts scalloped brown; juvenile and first-winter plumages have broad blackish postocular line and carpal bar.
Voice *Kre-ep* repeated, alarm *kit-kit-kit* while diving on intruders, and *zeet*.
Similar species All other white terns are larger, none has yellow bill with black tip and orange legs in breeding plumage.
Habitat and behaviour Sea coasts, lagoons, beaches, reservoirs. Flight swallow-like with fast wingbeats; hovers with head pointing downwards before plunge-diving for fish. Breeds May–July; builds nest on bare ground or sand; lays up to four speckled brown/blackish eggs.

Transition to non-breeding plumage. Note black tip to bill (eventually turning completely black), greyish-yellow legs and speckled crown.

Juvenile has dark scallops on back and crown.

Range *S. a. antillarum* breeds in eastern coastal United States, Texas to Honduras, and West Indies mainly in the Bahamas, Greater Antilles, Virgin Islands and Anguilla; winters in South America from the northern Pacific coast east to Brazil.
Status Fairly common summer breeding visitor, mainly on the cays, and passage migrant throughout the Cuban archipelago, March–early October. The population decline is due to loss of habitat and predation.

Bridled Tern

Onychoprion anaethetus 38cm

Breeding adult has white forehead extending behind eye, black crown, white collar and black bill.

Juvenile. Note mantle feathers edged white which create finely-barred effect.

Local name Gaviota Monja.
Taxonomy Polytypic (4).
Description Slender tern with dark grey mantle washed brownish, white underparts, black pointed bill and black legs. Breeding adult has white forehead extending behind eye, black loral line is same width from eye to bill, black crown and nape forming a 'bridle', narrow white collar. Non-breeding adult resembles breeding adult; may have paler streaked crown and nape and whitish lores. In flight shows mostly white underwing with dark-tipped primaries and trailing edge, long tail deeply forked, dark with broad white edges (mostly white undertail). Juvenile has streaked crown, dark brownish-grey upperparts with wavy bars, and shorter tail. First summer (March–September) has white head.
Voice Call soft *weep*, harsher *krrrrr* over nest-site.
Similar species Sooty Tern has white on forehead that does not extend behind eye, loral line narrows towards bill, and in flight underwing shows less white, with broader dark-tipped primaries.
Habitat and behaviour Marine sounds, cays and saline lagoons. Breeds colonially, May–August; builds nest in rock crevices, burrows under vegetation, or on sand; lays 1–2 bluish eggs marked with brown and lilac. Buoyant graceful flight, picking fish or squid from surface, seldom submerges.

Breeding adult in flight.

Range *O. a. melanopterus* breeds on islands off Caribbean coasts and locally in the West Indies in the Bahamas, Turks and Caicos Islands, Greater Antilles (except Jamaica), Virgin Islands and Anguilla. Winters at sea in the Atlantic–Caribbean region. Pantropical.
Status Common summer breeding visitor to northern and southern cays, April–September; otherwise rare.

Sooty Tern

Onychoprion fuscatus 38–43cm

Local name Gaviota Monja Prieta.
Taxonomy Polytypic (6).
Description Large black-and-white tern. Breeding adult has white on forehead extending only to eye, black loral line narrows towards bill, black cap, nape and mantle, black bill and legs. Non-breeding adult has blackish upperparts and may have black feathers with whitish edges. In flight underwing has whitish coverts, more extensive blackish on primaries and trailing edge, and deeply forked black tail with narrow white outer edge. Juvenile is black spotted with white on mantle and coverts.
Voice Usually silent; call *wide-a-wake*.
Similar species Bridled Tern has white forehead extending behind eye; only tips of feathers are blackish on underwing.
Habitat and behaviour Fringing reefs, coastal waters; makes shallow plunge-dives or lifts prey from surface. Breeds colonially, May–August on northern and southern cays; lays 1–3 white eggs, spotted and speckled with reddish-brown and pale lilac, on sand or bare rock.

Breeding adult in flight.

Breeding adult has white on forehead that extends to eye and loral line that narrows towards bill.

Range *O. f. fuscatus* breeds on the North American Atlantic–Gulf coast (irregularly), Mexico, and the West Indies, mainly in the Bahamas, Greater Antilles and Anguilla. Highly pelagic. The most abundant tern in the Caribbean. Pantropical.
Status Very common summer breeding visitor, though majority only seen on northern and southern cays, April–September.

Black Tern

Chlidonias niger 23–26cm

Local name Gaviotica Prieta.
Taxonomy Polytypic (2).
Description Small dark tern with small black bill, blackish patch extending onto side of upper breast in all plumages, short notched grey tail, long wings extend beyond tail at rest; in flight upperwing and underwing are grey. Non-breeding adult has blackish hindcrown extending from black ear-coverts, dark grey upperparts and blackish legs. Breeding adult black except for dark grey mantle and wings, white undertail-coverts and black legs. Juvenile has streaked crown, white scaling on brownish back, pinkish-red legs.
Voice Sharp *kyck*, usually silent.
Similar species Only small non-breeding tern with black ear-coverts joining blackish hindcrown and dark patch on sides of breast. Least Tern has yellow legs and deeply forked tail in all plumages.
Habitat and behaviour Coasts and wetlands. Buoyant flight with slow deep wingbeats on broad wings (like a nighthawk); frequently dips to take fish and insects from surface and may dive; perches on posts and roosts with waders.

First-winter.

Non-breeding adult in flight. Black cap continues to ear-coverts and black area on side of upper breast are diagnostic of this species.

Range *C. n. surinamensis* breeds in North America; winters on both coasts of Central America to both coasts of South America. Rare and irregular on passage throughout the West Indies. Cosmopolitan.

Status Uncommon to rare passage migrant on Cuba and a few northern cays, July–November and March–June.

Brown Noddy

Anous stolidus   38cm

Local name Gaviota Boba.
Taxonomy Polytypic (4).
Description Dark brown with blackish primaries and white-grey cap, darkening to brownish on nape. Long tail wedge-shaped and notched. Long bill and legs black. Juvenile and first-summer birds have whitish forehead and line above lores, rest of head brownish.

Voice Soft *kak* also, *carrrk*.
Similar species None.
Habitat and behaviour Open water near coasts and offshore cays. Flight is strong and erratic. Breeds May–August; crude nest low in trees or on the ground on bare sand or rock; lays 2–3 ashy-white eggs spotted with reddish-brown and pale lilac.

First-summer has white forecrown and supraloral.

First-summer. Note worn plumage that is visible on whitish upperwing-coverts.

Adult is entirely dark brown except for bright white cap that becomes grey towards nape.

Range *A. s. stolidus* breeds on islands in the Caribbean, Central America and southern Atlantic Ocean islands to the Gulf of Guinea. Pelagic during non-breeding season. Cosmopolitan.
Status Common local breeding resident on northern and southern cays, and in seas around Cuba.

Black Skimmer

Rhynchops niger   40–51cm

Local name Gaviota de Pico de Tijera.
Taxonomy Polytypic (3).
Description Unmistakable, with large red bill, black at tip, laterally compressed with mandible longer, thinner and projecting beyond maxilla; white forehead, underparts and outer tail (with dark central feathers); long narrow wings and short red legs. Non-breeding adult has dark brown upperparts, and white nuchal collar, duller bill and legs. Breeding adult has black upperparts. Juvenile has brownish-streaked crown and upperparts and blackish-red bill.
Voice Silent.
Similar species None.
Habitat and behaviour Beaches, fresh and saline lagoons, saltpans, coasts. Buoyant flight on very long pointed wings, with head held low; compact flocks fly up and down with sudden turns. Flies low over water with bill open and mandible submerged, ploughing water surface in search for small fish.

When foraging, adult holds wings above body and uses mandible to cut surface of the water.

Adult in flight. Upperwing shows contrasting white bar on trailing edge of long pointed black wings.

Breeding adults have black upperparts and a large red-and-black bill.

Range *R. n. niger* breeds on both coasts of United States to Mexico; winters from United States to both coasts of Panama and casually in the Bahamas, Greater Antilles and Cayman Islands.

Status Uncommon to locally common winter resident and passage migrant on Cuba and northern cays, July–April; numbers are increasing.

Scaly-naped Pigeon

Patagioenas squamosa 38cm

Local name Torcaza Cuellimorada.
Taxonomy Monotypic.
Description Adult is entirely dark grey with a wine-red hood, iridescent on hindneck; bare skin around eye is yellow to reddish-orange. Juvenile has brown head and neck.
Voice Resembles White-crowned Pigeon, but more mournful: *oooo, OO-oo-oo, ROO-OO,...OO-oo-oo, ROO-OO.*
Similar species White-crowned Pigeon has white or grey cap. Plain Pigeon has reddish-brown patch and white band on wing.
Habitat and behaviour Semi-deciduous and evergreen woodland, swamp woodland, tropical karstic woodland and rainforest, from sea level to high elevations. Breeding March–June. A frail stick nest built in base of palm fronds, or among bromeliad-laden branches; lays two glossy white eggs. Feeds on fruit, seeds and snails.

Adult.

Adult is entirely dark grey with a wine-red hood and iridescence on hindneck.

Adult in flight. Note bare skin around eye and dark unmarked wings.

Range Greater Antilles, except Jamaica, and the Lesser Antilles, as well as Curaçao, Bonaire and Los Testigos (off north-east Venezuela).
Status Common and local breeding resident on mainland Cuba, and several cays.

White-crowned Pigeon
Patagioenas leucocephala 33–36cm

Local name Torcaza Cabeciblanca.
Taxonomy Monotypic.
Description Large, entirely charcoal grey. Iris white, orbital skin pinkish-white, bill red tipped with grey, and legs red. Male has white crown, violet and green iridescence on nape and sides of neck giving a scaly effect, and raised brownish-red area on nape when breeding; female has crown greyish-white, reduced. Juvenile is entirely dark brown with back and wing-coverts edged buff. Immature has forecrown greyish, dull brownish-grey plumage, iris dark; adult plumage develops in second year.
Voice Deep throaty *croo cru* (rising) *cu-cruuu*, and soft rolling *cru-cruu*.
Similar species Scaly-naped Pigeon has wine-red head and lacks white crown. Plain Pigeon is brown, with white markings and reddish-brown patch on wing.
Habitat and behaviour Semi-deciduous and evergreen woodland, tropical karstic woodland, wetlands, rainforest. Frugivorous arboreal forager. Breeds April–August. Nests in colonies of many hundreds, generally on mangrove cays, also in woodlands. Builds rough platform nest of twigs, roots and plant stems 4–20m above ground; usually lays two white eggs; young are dependent on parents for up to 40 days.

Adult female has greyish-white crown.

Juvenile. Note dark brown plumage edged buff, and grey forecrown.

Male is charcoal grey with brilliant white crown and violet-green iridescence on nape.

Range Breeds on the Florida Keys, the Bahamas, Greater Antilles, Cayman Islands and Lesser Antilles as far south as Antigua; in decline throughout most of its range. Wanders widely between the Antillean islands.
Status Common to locally abundant breeding resident and passage migrant throughout the Cuban archipelago.

Plain Pigeon

Patagioenas inornata   38cm

Local name Torcaza Boba.
Taxonomy Monotypic.
Description Adult is brown with grey wings, rump and tail; upperwing shows thin white median band and reddish-brown patch; abdomen is vinaceous; iris white; eye-ring red. Juvenile duller with dark eye.
Voice A guttural call, *hoowua, HOO-hoowua, HOO-hoowua.*
Similar species Scaly-naped Pigeon has uniformly grey wings and wine-red head and neck. White-crowned Pigeon has white or grey cap.
Habitat and behaviour Coastal, semi-deciduous and evergreen woodland, low forested hills, grasslands with palm groves. Breeds April–July. Builds a fragile stick nest lined with grass; lays two white eggs. Feeds on fruit, seeds.

Adult. Note reddish-brown patch and thin white median band on upperwing. Iris is white and eye-ring red.

Adult.

Antillean endemic.

Antillean endemic.

Range Endemic to the Greater Antilles.
Status Near Threatened. Rare breeding resident, very local on mainland Cuba, where remaining populations are in coastal areas of Guanahacabibes Peninsula; Sierra de Najasa, Camagüey province; Birama, Granma province; and the Isle of Pines. Also recorded in some northern and two southern cays, possibly stragglers.

Eurasian Collared-Dove
Streptopelia decaocto 30cm

Adult.

Local name Tórtola de Collar.
Taxonomy Polytypic (2).
Description Entirely pale greyish-brown with incomplete black collar. Tail long and square with slightly rounded corners, tipped greyish-white; grey undertail-coverts.
Voice Soft, repeated *kuuk-kooooooo-kuuk*.
Similar species None.
Habitat and behaviour Urban and rural areas. Feeds on fruit and grain. Breeds March–December, possibly year-round. Nest is an unlined platform of twigs in a bush or tree; lays two white eggs.

Adult. Note incomplete black collar, and long square tail with white tip.

Range Throughout Eurasia. Introduced in the Bahamas in 1974, and has subsequently invaded Florida, Cuba and much of Lesser Antilles.
Status Confirmed on Cuba in 1990. Currently widespread but local breeding resident on Cuba and some northern cays.

White-winged Dove

Zenaida asiatica 28–30cm

Local name Paloma Aliblanca.
Taxonomy Polytypic (3).
Description Adult brownish-cinnamon with distinct broad white band on central wing. Wide blue orbital ring and orange iris, black crescent under cheek; sides of lower neck have iridescent purple sheen, usually more conspicuous in male. Bill black and feet red; outer tail feathers broadly tipped white. Juvenile pale and greyish, iridescence absent, iris dark.
Voice Perhaps the loudest among pigeons: *ooh-woo-woo-woo*. Two common rhythms are *ooah-AHoo* and *who cooks for you all* or *Cru-cu-ca-roo* on level pitch.
Similar species Zenaida Dove is smaller with narrow white band on trailing edge of secondaries.
Habitat and behaviour Semi-deciduous and evergreen woodland, pine forests, tropical karstic woodland, coastal thickets, wetlands. Fast and high, direct flight. Large flocks form where food is plentiful. Forages in trees and sometimes on ground, especially

Adult has black streak below eye, blue orbital ring, orange iris and white band along wing.

in urban areas. Breeds April–July. Sparse twig nest, from 1.3–20m height; lays two white eggs.

Adult in flight shows white wing-coverts above, grey underwing-coverts and white terminal tail-band.

Range *Z. a. asiatica* breeds in southern United States from Florida through most of Middle America and the West Indies in the Bahamas, Greater Antilles, San Andrés and Providencia in the western Caribbean. Northern birds migrate south in winter.
Status Common breeding resident throughout the Cuban archipelago.

Zenaida Dove

Zenaida aurita 28cm

Local name Paloma Sanjuanera, Guanaro.
Taxonomy Polytypic (3).
Description Cinnamon-brown above, underparts vinaceous-brown, distinct white band on trailing edge of secondaries very conspicuous in flight. Two black ear-covert spots, violet blue-orbital ring and dark iris; sides of lower neck iridescent purple, usually more conspicuous in male, with black spots on wing-coverts; legs red. Tail rounded with black subterminal band and light grey tips to outer rectrices. Juvenile has buff-white edges to back and wing-coverts and lacks iridescence.
Voice A clear *OOLA, OO, OO-OO*, with a rather sharp second syllable.
Similar species Mourning Dove has long, pointed tail and lacks white tips to secondaries, and has erratic flight. White-winged Dove is larger with broad long white band on wing.
Habitat and behaviour Open woodland, open country with trees, semi-deciduous and evergreen woodland, swamp woodland, second growth, coastal thickets, pine forest and clearings. Forages mainly on the ground for seeds, also fruits in trees. Breeds March–July. Single pairs usually defend territories; builds a thin platform of twigs in a bush or tree; lays two white eggs.

Adult. Note iridescent purple on hindneck, black mark above and below ear-coverts, and large black spots on coverts.

Adult. In flight displays white trailing edge to secondaries and dark primaries.

Range Northern Yucatán coast and islands offshore, and the West Indies. *Z. a. zenaida* is resident in the Bahamas, Greater Antilles to the Virgin Islands, and Cayman Islands.
Status Common breeding resident throughout the Cuban archipelago.

Mourning Dove
Zenaida macroura 28–33cm

Local name Paloma Rabiche.
Taxonomy Polytypic (5).
Description Greyish-brown upperparts, paler underparts and long, pointed tail bordered in black and white. Male has sides of lower neck iridescent purple; female paler with iridescence reduced. Both have single ear-covert spot and black spots on wing. Juvenile is grey-brown with back and wing-coverts edged buffy, buffy ear-coverts and faint spotting on breast.
Voice Lengthy and mournful *OOWA, OO, OO-OO*. Very similar to Zenaida Dove, but slightly more drawn out, breathy, and higher pitched; second syllable not as clipped.
Similar species Zenaida and White-winged Doves have shorter, rounded tail with white markings on wing.
Habitat and behaviour Grasslands, agricultural land, wetlands, open semi-deciduous, evergreen and swamp woodland, coastal thickets, pine forest, second growth, urban areas. Feeds on seeds, grain, some fruit

Adult. Note black spot on ear-coverts.

and snails; sometimes forms large flocks where food is abundant. Flight nervous and quick, with irregular wingbeats. Breeds February–September in flimsy nest; lays two white eggs.

Adult. Note spots on wings and long pointed tail.

Range *Z. m. macroura* breeds in Florida Keys, Cuba, Isle of Pines, Jamaica, Hispaniola and Puerto Rico. *Z. m. carolinensis* breeds from south-east Canada to southern United States, Bermuda and Bahamas; winters south throughout most of the breeding range.
Status Abundant breeding resident throughout the Cuban archipelago.

Common Ground-Dove
Columbina passerina 15–18cm

Adult. Note rufous on primaries and underwing-coverts when wings are raised.

Female is duller than male with sandy-grey scaling on head and breast.

Local name Tojosa.
Taxonomy Polytypic (18).
Description Very small; greyish-brown, with faint scaling on head and breast and scattered, small dark violet spots on wings. In flight, primaries and wing linings are rufous, tail short and rounded with black edges and white corners. Male has bluish-grey crown and upperparts, pinkish-cinnamon forehead, face and throat, beautifully marked on breast and sides of neck with pinkish pale grey feathers darkly edged, giving a unique scaling effect; pink bill with black tip; legs pink and eye orange. Female is slightly duller and crown is sandy-grey. Juvenile similar but back and wing-coverts edged whitish; scaly head and greyish breast; bill dark.
Voice Call steady, level single *coo coo coo coo* or *co-coo co-coo co-coo* and whooping *hoap-hoap-hoap*.
Similar species Only diminutive dove.
Habitat and behaviour Swamp, semi-deciduous and evergreen woodland borders, wetlands, grassland second growth, agricultural land, coastal thickets and urban areas. Breeds January–July; pair bonds retained throughout year. Builds thin, frail nest of fine twigs, grasses, stems, rootlets and occasionally feathers; lays two white eggs; young fed by both parents and remain in family group. Also in foraging flocks of up 15 individuals. Flies only short distances when disturbed, vigorously bobs its head while walking.

Adult male has blue and pink scaling on head and breast and iridescent blue spots on wing-coverts.

Range Breeds from the southern United States through Middle America to Ecuador and Brazil; also, islands in the western Caribbean and throughout the West Indies. *C. p. insularis* occurs on Cuba, Cayman Islands and Hispaniola and associated Islands.
Status Common to abundant resident throughout the Cuban archipelago.

Key West Quail-Dove
Geotrygon chrysia 28cm

Local name Barbiquejo.
Taxonomy Monotypic.
Description Adult male is reddish-brown above, breast pale vinaceous and belly pale grey; purple-green iridescence on head and back; rufous wings; tail bright dark rufous. Conspicuous white stripe under eye; red iris; bill pinkish at base and horn-coloured tip; pinkish legs. Sexes similar but female duller and less iridescent. Juvenile almost without iridescence, pale borders on wing-coverts, darker on upper breast and side of neck.
Voice Very deep *ooowooo*, sliding slightly downward in pitch and repeated in series at *c.* 3-second intervals. It is quite similar to Ruddy Quail-Dove, but slower, and more mournful.
Similar species Male Ruddy Quail-Dove is smaller with reddish-brown back, pinkish-brown underparts, buff stripe under eye; female is olivaceous-brown.
Habitat and behaviour Coastal thickets, swamp, semi-deciduous and evergreen woodland. Breeds September–July. Builds nest near the ground or low in trees, laying two beige eggs. Solitary outside breeding season. Feeds on fruit, seeds, small snails; usually on ground; also perches on trees low to mid-level.

Male has purple iridescence on back and coverts, green on head and nape, and rufous on wings and tail.

Female is duller than male, being reddish-brown with less iridescence.

Range Bahamas, Cuba, Hispaniola (including Gonâve, Tortue and Catalina islands), Puerto Rico and Vieques Island.
Status Uncommon to locally common breeding resident on Cuba, Isle of Pines, northern cays and Cayo Cantiles off south coast.

Gray-fronted Quail-Dove

Geotrygon caniceps 28cm

Local name Camao, Azulona.
Taxonomy Monotypic.
Description The only quail-dove with white and greyish head, iris red, bill flesh-coloured at base and ivory tip. Conspicuous blue and purple iridescence on back and side of breast; lower back and rump bluish-grey; tail grey and legs pinkish. Underparts grey with rufous from mid-belly to undertail-coverts. Female slightly duller, greyer on head with less iridescence. Juvenile brownish.

Adult.

Voice Low, rapidly repeated *uup-uup-uup* in a long series, and more measured, louder and ascending *uoop*.
Similar species None.
Habitat and behaviour Swamp forest, semi-deciduous and evergreen woodland, rainforest; from sea level to high elevations. Breeds January–August; loose nest of twigs and leaves, 1–3m above the ground, on the main trunk or a branch; single cream-coloured egg, rarely two. Feeds on fruits and seeds. Bobs body after perching or while walking.

Juvenile.

Adult. White-grey head is diagnostic among quail-doves on Cuba. Note purple iridescence on back.

Range Widespread but local on Cuba.
Status Vulnerable. Endemic species, rare, confined to mainland Cuba, most common on Zapata Peninsula.

During the drought season, Gray-fronted Quail-Doves look for water holes in the forest to drink from.

Adult.

Ruddy Quail-Dove

Geotrygon montana 25cm

Local name Boyero.
Taxonomy Polytypic (2).
Description Male reddish-brown above, back washed with purple; pronounced buff stripe below eye; underparts vinaceous-buff. Female has olivaceous-brown upperparts and beige-brown underparts, except for white throat; facial stripe less distinct than in male. Juvenile similar to female, back feathers tipped with ochre.
Voice A simple, deep *hooo* with no internal variation in pitch, repeated in series at *c.* 3-second intervals.
Similar species Key West Quail-Dove is larger with green and purple iridescence on head and back, and greyish underparts.
Habitat and behaviour Dense semi-deciduous and evergreen woodland, swamp, tropical karstic woodland; from sea level to medium elevations. Breeds January–July; loose nest of twigs and leaves 1–6m above ground, lays two cream-coloured eggs. Feeds on seeds, fruit, small snails. Usually perches low, and like other quail-doves most often found on ground. Flight low and very fast; flushes more readily than other quail-doves.

Juvenile is olive-brown with pale feather edges on back and crown.

Male is reddish-brown above, with back washed purple, and a buff stripe below eye.

Female duller and greyer than male and has indistinct facial stripe.

Range Middle America south to northern Argentina; West Indies.
Status Common breeding resident on Cuba, Isle of Pines and some larger northern cays.

Blue-headed Quail-Dove

Starnoenas cyanocephala   30cm

Local name Paloma Perdiz.
Taxonomy Monotypic.
Description Cap bright blue, sides of neck streaked, and long, conspicuous white stripe under brown eye. Cinnamon-brown with olive wash, apart from bib of black throat and breast flecked with iridescent blue and bordered with white; reddish-purple legs. Juvenile similar, with dark-tipped cap feathers.
Voice A deep two-syllable *oooowup...oooowup*, repeated in long series at *c.* 2-second intervals, each note ending abruptly.
Similar species Key West Quail-Dove has green and purple iridescence on head and pale vinaceous breast.
Habitat and behaviour Semi-deciduous, evergreen and swamp woodland, second growth with abundant leaf litter, on limestone; commonest at sea level, but also occurs in cloud forest at higher elevations (up to 1,792m). Breeds March–June. Loose nest of twigs, among bromeliads or on trunk near ground; lays two white eggs. Solitary or in family groups. When artificially fed, forms flocks of up to 15 individuals.

Adult. Note metallic blue crown and long white stripe under eye.

Adult baving a sun bath.

Range Widespread on Cuban mainland, but local. Virtually extirpated on Isle of Pines.
Status Endangered. Endemic species, uncommon breeding resident, **vulnerable** on mainland Cuba.

Pair of Blue-headed Quail-Doves.

Adult displaying.

Cuban Parakeet

Psittacara euops 25cm

Local name Catey.
Taxonomy Monotypic.
Description Vivid green, with scattered red spots on head, sides of neck and breast (number of spots variable). A large red area at bend of wing shows in flight as red coverts, with golden underwing; tail long and pointed. Juvenile duller.
Voice Repeated squeak *crik-crik-crik*, mostly in flight; a low murmur while perched.
Similar species None.
Habitat and behaviour Woodlands, grasslands with palm groves. Breeds April–August. Lays 2–5 white eggs in abandoned woodpecker holes in palms, and occasionally in natural holes in cliffs. Feeds on seeds and fruits.

Adult in flight. Note red on lower face, throat and abdomen; tail yellowish-green with red at base.

Group allopreening. Note red at bend of wing and greenish-yellow undertail-coverts.

Range Cuba.
Status Vulnerable. Endemic species, local and rare. Restricted to Zapata Peninsula, mountains of Trinidad, Loma de Cunagua, Sierra de Najasa and some areas of eastern mountain ranges.

Adult.

Adult.

Adult. Note green plumage, bare skin around eye and long pointed tail.

Cuban Parrot

Amazona leucocephala 28–33cm

Local name Cotorra.
Taxonomy Polytypic (4).
Description Bright green plumage, feathers edged blackish, white forehead and facial skin around eye; blackish ear-coverts, red cheeks, chin and throat, darker red on abdomen; wide creamy-whitish thick pale bill with deeply curved culmen, pale legs and feet. Blue band on closed wing shows as blue primaries in flight, undertail-coverts yellowish-green, tail green with basal part red, tip yellowish. Wingbeats shallow and rapid.
Voice Detected usually by calls. Very vocal with unmusical whistles, bugles and squawks, especially when preparing to leave the roost. Wide range of calls for flight, contact, territory and threat; flight call a hoarse *d-de*; also, conversational murmur among feeding groups.
Similar species None.
Habitat and behaviour Semi-deciduous and evergreen woodland, swamp woodland, rainforest, grassland with palm groves. Breeds February–July. Lays 3–4 round white eggs in abandoned woodpecker holes, usually in palms; female only broods for *c.* 28 days and both parents feed young; 3–4 young fledge after 53–55 days and beg from parents for several weeks, remaining in small family groups. Feeds on seeds and fruits.

Pair at nesting cavity (female on left; male on right). Both sexes have white forehead and eye-ring.

Adult. Note green plumage, and red cheeks, chin and throat.

Adult in flight. Note red cheek, throat and abdomen; tail yellowish-green, with red at base.

Range West Indian endemic. *A. leucocephala* is resident on Cuba, Bahamas and Cayman Islands.
Status Near Threatened. Endemic subspecies *A. l. leucocephala* is locally common breeding resident on Cuba, restricted to Guanahacabibes Peninsula, Zapata Peninsula, southern and western Isle of Pines, Macizo de Guamuhaya, Loma de Cunagua, Sierra de Najasa, the forests of western Sierra Maestra and Cuchillas del Toa.

Black-billed Cuckoo
Coccyzus erythropthalmus   30cm

This individual shows buff throat, white underparts, black bill and a red orbital ring.

Local name Primavera de Pico Negro.
Taxonomy Monotypic.
Description Long slender bird with grey-brown upperparts, red eye-ring and thin, decurved blackish bill; whitish underparts and long tail with small white tips. In flight may show little or no rufous on primaries. Juvenile with upperwing-coverts edged whitish or buff, buff wash on throat; eye-ring grey, changing in autumn from buff to yellow.
Voice A low, hollow *cu-cu, cu-cu-cu* or *cu-cu-cu-cu*, repeated rhythmically in long series, mostly on breeding grounds.
Similar species Yellow-billed Cuckoo has thick bill with yellow mandible, large white tail-spots and distinct, bright rufous primaries. Mangrove Cuckoo has black ear-patch and buffy-cinnamon underparts.
Habitat and behaviour Coastal thickets.

A bird in transition to full adult plumage.

Range Breeds in North America; winters in north-western South America; rare on passage in the West Indies in the northern Bahamas, Cuba and Hispaniola, vagrant elsewhere.

Status Very rare but easily overlooked passage migrant on Cuba, Isle of Pines and some cays, September–November and March–May.

Yellow-billed Cuckoo

Coccyzus americanus 28–32cm

Local name Primavera de Pico Amarillo.
Taxonomy Polytypic (2).
Description Long slender bird with greyish-brown upperparts, white cheek and underparts, and rufous primaries on folded wing. Stout bill slightly downcurved, almost entirely yellow except for dark central culmen and tip; grey or yellow orbital ring; narrow rounded tail with black outer retrices tipped with broad white markings, mostly visible from below. In flight shows rufous primaries and large white spots on edges of long tail feathers. Juvenile with orbital ring yellow and paler undertail pattern.
Voice Fast, guttural cackling, diminishing in loudness *ka-ka-ka-ka-ka-ka, kow-kow-kow*.
Similar species Mangrove Cuckoo has black patch behind eye, buffy underparts and, in flight, brown wings. Black-billed Cuckoo (very rare) has grey, thinner bill and small tail-spots, wings show little or no rufous.
Habitat and behaviour Semi-deciduous and evergreen woodland, swamp woodland, coastal thickets. Breeds April–July. Nest a shallow twig platform, with rootlets, dry leaves and grasses, in a bush or tree; lays 2–5 greenish-blue eggs. Occasionally parasitises nests of other birds. Typical fast low flight. Feeds on insects, caterpillars, lizards.

Adult has white underparts; underside of closed tail shows white spots and tip.

Adult. Note rufous wing-patch that is even more obvious in flight.

Range South-eastern Canada, United States, Mexico and Greater Antilles, wintering from northern South America east of the Andes to northern Argentina.
Status Uncommon summer breeding visitor and passage migrant throughout the Cuban archipelago, February–December.

Mangrove Cuckoo

Coccyzus minor 28–30cm

Local name Arrierito.
Taxonomy Monotypic.
Description Long slender bird with greyish-brown crown and hazel-brown upperparts, black patch from lores over eye to ear-coverts, yellow orbital ring, long heavy bill downcurved with most of mandible yellow with dark tip; white throat, breast pale buffy-cinnamon, becoming richer on belly and undertail-coverts; very long tail with large white spots on black outer retrices. In gliding flight shows unmarked wings (no rufous) and tail with broad white spots. Juvenile lacks black on face, eye-ring grey, back and wing-coverts edged cinnamon, buffy-cinnamon underparts and less distinct tail-spots.
Voice Unmistakable long nasal series *ge-ge-ge… gou-gou-goul-goul,* slower at end.
Similar species Yellow-billed Cuckoo and Black-billed Cuckoo have white underparts and lack black ear-coverts.
Habitat and behaviour Coastal thickets, mangroves. Inactive and approachable; widely distributed. Forages for insects and spiders slowly at mid–low heights. Breeds April–July. Nest twig platform often below 3m elevation; lays three greenish-blue eggs.

Adult has black band over eye to ear-coverts, long bill with most of mandible yellow, pale buffy-cinnamon underparts and large white spots on tail.

Range Breeds in southern Florida, Middle America to northern South America, and the West Indies; winters throughout breeding range.
Status Uncommon breeding resident, rare in central and western Cuba, less so in eastern Cuba, and on many larger cays.

Great Lizard-Cuckoo
Coccyzus merlini 51cm

Local name Arriero.
Taxonomy Polytypic (4).
Description Large. Olive-brown upperparts; grey-buff throat and breast with reddish-brown belly. Tail very long, undertail has bold pattern of black and white. Bill blue-grey, very long, slightly decurved. Bare skin around eye orange-red. In flight, shows broad rufous wings and outer tail tipped black and white. Three subspecies have been described for the Cuban archipelago. Those from the Isle of Pines and northern cays are smaller than the main-island race and have paler, more extensive grey on underparts. Juveniles of all subspecies have paler tail pattern and yellow bare skin around eye.
Voice A two-part call, *tacooo-tacooo, ka-ka-ka-ka-ka-ka*, the second part gradually becoming louder; also a guttural *tuc-wuuuh*.
Similar species Mangrove and Yellow-billed Cuckoos are considerably smaller, with shorter tails and bills.

Cuban mainland subspecies *S. m. merlini*. Grey and buff below, tail very long, undertail has a bold black-and-white pattern.

Cuban mainland subspecies *S. m. merlini*. Note the very long, slightly decurved blue-grey bill, and bare orange-red skin around eye.

Habitat and behaviour Semi-deciduous and evergreen woodland, from sea level to fairly high elevations, coastal thickets, tropical karstic woodland, swamp woodland. Breeds April–October. Flimsy stick nest with leaves, at low or middle levels in trees or bushes; lays 2–3 white eggs. A poor flier; commonly flaps and glides from treetops and often forages as it walks or runs on ground or among branches. Feeds on frogs, snakes, birds' eggs and nestlings, large insects.

Northern cays subspecies *C. m. santamariae* has whiter underparts.

In flight adult displays bright rufous wings and long white and black-tipped tail.

Range Bahamas and Cuba. Three subspecies on Cuban archipelago *C. merlini merlini* (mainland), *C. m. decolor* (Isle of Pines), *C. m. santamariae* (Cayo Santa María, Cayo Coco).
Status Common breeding resident on Cuba, Isle of Pines and some northern cays.

Smooth-billed Ani

Crotophaga ani 30–33cm

Adult has keel-like ridge on curved culmen and very long tail.

Local name Judío.
Taxonomy Monotypic.
Description Adult has blue-black plumage with iridescent bronze sheen on back and wings, large black bill with ridged, curved culmen projecting to crown, very long tail; head and neck feathers are short and scaly with bare skin around eye. Juvenile is smaller with brownish-black feathers edged grey.
Voice Rising whistled *keuu-iik*, and loudly repeated low *ann-ee*; also whistles and harsh squawks.
Similar species None, heavy curved bill is diagnostic.
Habitat and behaviour Grassland, agricultural land, open areas in swamp woodland, tropical karstic woodland, second growth. Small flocks usual in disturbed habitats with grassland. Forages for arthropods, also in trees taking young birds, lizards, snakes and fruits. Laboured flap-and-glide flight on short rounded wings. Breeds April–October. Bulky communal nests are shared with several females, 4–30 bluish eggs with flaky white coating laid in layers separated by leaves; only the top layer hatches.

Juveniles. Note smaller and less-grooved bill and brownish-black plumage.

Range Breeds in Florida, Costa Rica to northern Argentina; West Indies.
Status Very common breeding resident throughout the Cuban archipelago.

Barn Owl

Tyto alba 29–43cm

Local name Lechuza.
Taxonomy Polytypic (24).
Description Appears entirely white at night. Adult has large rounded head, heart-shaped white facial disc with large dark eyes, body tapers to short tail and feathered legs with well-developed talons. Upperparts flecked orange and gold with greyish spots, underparts and underwing whitish and buff with flecks of brown. Plumage and long broad wings adapted for silent flight. Sexes similar but female larger and darker. Juvenile has dark brown upperparts and face and is more spotted and streaked.
Voice A raspy hissing scream *shhh*, hisses, squeaks and clicks, a shriek *shreeeeeeee* heard while hunting, a two-syllable screech during breeding displays and a variety of noises at nest; also, young birds emit constant calls.
Similar species None.
Habitat and behaviour All habitats with suitable nesting and roosting sites: cavities in palms or *Ceiba* trees, caves, lofts of houses, church towers, abandoned buildings. Mainly a nocturnal and crepuscular hunter, but also observed in mid-morning or mid-afternoon on dull cloudy days. Breeds year-round, peaks December–May, preceded by noisy pair-bonding displays. Highly territorial at nest-site. Lays 2–5 white eggs in natural cavities or abandoned buildings; incubated by female, fed by male. Young fledge after about 10–11 weeks, often remaining with parents for several months. Second brood is raised in optimum years. Feeds mainly on rodents; also reptiles and small birds.

Juvenile is browner overall, including facial disc.

Female. Note flecked orange and gold underparts.

Adult in flight.

Range Breeds throughout the Americas and West Indies where *T. a. furcata* is confined to Cuba, Jamaica and Cayman Islands. Cosmopolitan.
Status Common breeding resident on Cuba, Isle of Pines and some northern cays. North American *T. a. pratincola* is possibly a vagrant or rare winter visitor.

Bare-legged Owl

Margarobyas lawrencii　22cm

Adult.

Adult in characteristic pose from tree hollow.

Local name Sijú Cotunto.
Taxonomy Monotypic.
Description Small owl, dark brown upperparts mottled with white; pale greyish-white underparts with faintly streaked breast and belly. Conspicuous white supercilium, large brown eyes, and pale legs long and unfeathered. Strictly nocturnal.
Voice A series of 12–15 low *cu* notes, rapidly accelerating in the manner of a bouncing ball. Pairs sing antiphonally on slightly different pitches. Also, a repeated *wheer* or *wheep*, rising and then falling.
Similar species None.
Habitat and behaviour Semi-deciduous and evergreen woodland, swamp woodland, coastal thickets, tropical karstic woodland. Breeds March–May. Nests in natural cracks and holes in trees, especially palm stumps or old woodpecker nests; two white eggs usual. Feeds on small reptiles, birds, large insects.

Range Cuba.
Status Endemic species. Common on Cuba, Isle of Pines and some northern cays such as Cayo Coco and Cayo Romano.

Adult. Note very large, dark-brown eyes with wide buffy stripe above, white spots on back, and bare legs.

Cuban Pygmy-Owl
Glaucidium siju 17cm

Local name Sijú Platanero.
Taxonomy Polytypic (2).
Description Smallest owl, compact with yellow eyes. Brown upperparts with dark spots on back of head resemble eyes; tawny upper breast more pronounced in female; underparts white, mottled with brown; short legs are entirely feathered. Juvenile has finely streaked abdomen. Occurs as two morphs, brown and grey.
Voice A single, rather high-pitched *too* at intervals; an accelerating series of *kew* notes, rising in pitch and changing in quality. Also, a high-pitched *whee*, heard commonly during courtship.
Similar species None.
Habitat and behaviour Semi-deciduous and evergreen woodland, swamp woodland, pine forest, rainforest and tropical karstic woodland from sea level to high mountains. May cock tail while perched, and jerkily moves it from side to side. A diurnal and nocturnal hunter; usually easy to approach. Breeds December–May. Nests in natural cavities or abandoned woodpecker holes; lays 2–4 white eggs. Feeds on lizards, frogs, snakes, insects, small birds, mice.

Adult.

Adult has round head and yellow eyes, feathered legs and short banded tail.

Range Cuba.
Status Endemic species with two subspecies: *G.s. siju* is a common breeding resident on Cuba, Cayo Romano and Cayo Cantiles. *G. s. vittatum* is common on the Isle of Pines.

Adult.

Adult.

Dark spots on back of head resemble eyes.

Burrowing Owl

Athene cunicularia 23cm

Local name Sijú de Sabana.
Taxonomy Polytypic (20).
Description Dark brown upperparts extensively and conspicuously spotted with white. Eyes yellow. Breast and belly whitish, barred with brown; white stripe across chin. Legs long; tail short. Juvenile has underparts buff and unbarred.
Voice A rapid, chattering *quick-quick*.
Similar species Bare-legged Owl is similar in size but is a strictly nocturnal forest species with brown eyes and faint streaks below.
Habitat and behaviour Grassy-sandy savannas, grassland, rocky fields. Terrestrial, hunts by hovering over open country, frequently in daylight. Frequently bobs up and down when agitated. Breeds May–August. Lays 5–8 round white eggs at the end of excavated tunnels up to 2m long. Feeds on small reptiles, large insects.

Adult, breeding resident population.

Adult shows barred underparts.

Range Western and central North America and Florida, central Mexico. Locally in South America, Bahamas, Cuba, Hispaniola (including Gonâve and Beata islands), and northern Lesser Antilles.
Status Rare breeding resident and local subspecies *A. c. guantanamensis* from eastern Cuba; western and central populations have undefined subspecific status. The migratory *A. c. floridana* is possibly a rare winter visitor recorded mainly from some cays.

Stygian Owl

Asio stygius   43cm

Local name Siguapa.
Taxonomy Polytypic (6).
Description Large, dark brown; upperparts spotted with cream and white, underparts streaked, and two long ear-tufts. Facial disc brown. Eyes yellow-orange. Juvenile is grey and entirely barred.
Voice Male, a short, hushed, *hooo*; female, *quick-quick*.
Similar species Short-eared Owl is smaller, with barely visible ear-tufts and pale buff facial disc, and occurs in open country.
Habitat and behaviour Semi-deciduous and evergreen woodland, rainforest, swamp woodland, pine forest. Breeds November–April. Nests high in trees; may add sticks to an old hawk nest or build a stick platform; lays two white eggs. Feeds on birds, bats and insects.

Adult.

Juvenile.

Adult is very dark brown with conspicuous ear-tufts, orange eyes and dark heavy streaks on underparts.

Range Mexico, Central and South America; Cuba, Hispaniola and Gonâve.
Status An endemic subspecies, *A. s. siguapa* is an uncommon breeding resident on Cuba, Isle of Pines and some northern cays.

Short-eared Owl

Asio flammeus 35–43cm

Local name Cárabo.
Taxonomy Polytypic (11).
Description Head large with round, brownish facial disc, black patches around large yellow iris and contrasting white feathering around bill. Plumage cryptic with upperparts dark brown spotted buff, breast buff with heavy streaks, tail barred with white terminal band, short black bill, legs feathered. Female larger, darker and more heavily streaked. Juvenile has dark face, browner upperparts edged buff and fainter streaks below. Buoyant flight on long, broad, pointed wings; underwing shows large yellowish-buff patch at base of outer primaries contrasting with black carpal patches and black tips to primaries.
Voice Hissing, wing clapping and bill snapping. Call *beee-ow*; and *kuk-kuk-kuk*.
Similar species Stygian Owl is bulkier, a forest bird with long ear-tufts, and tawny-brown facial disc.
Habitat and behaviour Grassland and agricultural lands. Active hunter during the day, night, dawn and dusk, quartering slowly over land for lizards, rats, bats, birds and insects detected by sight and sound; also roadkills. Roosts and breeds year-round on ground in grassland, where two nests lined with grasses and feathers had 3–4 white eggs.

Adult has round facial disc, rimmed in white, dark around yellow eyes, and feathered legs.

Adult in flight. Note black-tipped primaries and carpal patch on wings, and barred wings and tail.

Range Breeds in North and South America, and western Greater Antilles. The endemic subspecies *A. f. cubensis* breeds on Cuba. Northern North American subspecies *A. f. flammeus* winters to southern United States, Mexico and Greater Antilles. Cosmopolitan.
Status Common breeding resident on Cuba, increasing in numbers, and some northern cays. Not known from Isle of Pines.

Common Nighthawk

Chordeiles minor 22–25cm

Local name Querequeté Americano.
Taxonomy Polytypic (9).
Description Large flat head and small bill with a wide gape. Mottled dark greyish-brown, underparts strongly barred whitish-grey. Wings extend beyond tail at rest. In flight long, blackish pointed wings are bent in a V at the carpal joint, conspicuous white band across the primaries, tail slightly forked and barred black and white. Small weak legs. Male has white throat patch, broad white band on five outer primaries and broad white subterminal band on tail. Female throat patch smaller and buffy, less distinct white band on primaries, underparts buffy rather than greyish, white subterminal band absent. Immature similar to female but underparts heavily barred.
Voice Two-syllable call *pee-nt,* seldom heard on migration.
Similar species Antillean Nighthawk is very similar but wing-linings and underparts have more buff; rapid wingbeats with shorter glides between; best differentiated by call.

Habitat and behaviour Open grassland, agricultural and coastal areas. Diurnal and crepuscular, foraging for insects with fast erratic bounding flight and deep wingbeats interspersed by long glides.

Adult.

Adult. Whitish-grey underparts. Almost identical to Antillean Nighthawk; in flight they can be safely separated by their calls.

Range Breeds from North to Middle America; winters in South America. The majority migrate through Middle America; uncommon on passage in the Greater Antilles and Cayman Islands.
Status Uncommon passage migrant on Cuba and Isle of Pines, August–October and April–May.

Antillean Nighthawk
Chordeiles gundlachii 20–25cm

Local name Quereequeté.
Taxonomy Monotypic.
Description Adult has upperparts blackish-brown speckled whitish-grey, cinnamon, russet, black and tawny; underparts and wing-linings strongly barred and suffused with buff. Large flat head, large brown eyes, small bill with a wide gape and small, weak legs. Wings regularly do not extend beyond tail at rest, and in flight are long, slender and pointed, bent in a V at the carpal joint. Male has large white throat patch; in flight shows broad white band on five outer primaries and broad white subterminal tail-band. Female has buff throat and lacks tail-band. Juvenile similar to female, underparts heavily barred, pale panel on the upperwing, primaries and secondaries tipped with white.
Voice In flight gives diagnostic constant three-to-five-syllable call *rickery-dick* or *ke-re-ke-te* with slight accent on last syllable.
Similar species Common Nighthawk is virtually identical, is more greyish on wing-linings and underparts, wings extend beyond tail at rest but safe identification depends on hearing calls.
Habitat and behaviour Swamp woodland, pine forest, second growth and urban areas. Mostly crepuscular (also seen or heard at night and on cloudy days). Feeds on insects caught on wing, singly, in pairs or large flocks. Breeding April–July. Male courtship includes 'booming' in a series of steep dives and displaying fanned tail. Lays single (rarely two) bluish-brown spotted egg on bare ground, incubated by female. Perches lengthways on tree branches; bounding flight consists of deep wingbeats interspersed with long glides.

Male fanning tail during courtship.

Adult female lacks white throat patch.

Adult male has white throat patch and tail spots that form a white subterminal band in flight.

Male in flight displays white subterminal tail band that is absent in female.

Range *C. g. gundlachii* breeds in the Bahamas, Greater Antilles, Cayman Islands and southern Florida; it is thought to winter in South America.
Status Common summer breeding visitor and passage migrant throughout the Cuban archipelago, February–October.

Chuck-will's-widow

Antrostomus carolinensis 31cm

Local name Guabairo Americano.
Taxonomy Monotypic.
Description Adult has overall cryptic coloration, mottled brown with whitish scapulars tipped black, whitish band around throat, unspotted breast; small bill with huge mouth and long rictal bristles. Perched, wing reaches more than halfway along tail. Male larger, with white inner webs to three outer tail feathers and buff terminal band. Female has all-brownish tail with buff terminal band.
Voice A four-syllable *chuk-will-wi-dow*; a croaking sound given in flight.
Similar species Cuban Nightjar has more spotted underparts; the tail is tipped whitish in males.
Habitat and behaviour Nocturnal, feeding on insects caught on wing from a perch or by hawking. Roosts during the day in trees or on the ground, from which it is not easily disturbed. Flight fast, alternating wingbeats with glides.

Adult.

Adult. Note flattened head, cryptic plumage and white scapulars spotted black.

Range Breeds in eastern North America. Winters in southern United States, Middle America to northern South America and Bahamas and Greater Antilles (common on Hispaniola and Cuba, uncommon to rare elsewhere); on passage in the Cayman Islands.
Status Fairly common in winter and on passage throughout the Cuban archipelago, August–May. Possibly breeding, no data.

Cuban Nightjar

Antrostomus cubanensis 28cm

Local name Guabairo.

Taxonomy Polytypic (2).

Description Mottled greyish-brown upperparts, with breast and belly blackish-brown spotted with white. Bristles very long and curved inward; bar at base of throat buffy, tail and wings rounded. Male has conspicuous white tips on outer tail feathers, buffy in female.

Voice A repeated *tu-wurrrr*, often preceded by a quick *tuk* audible only at close range. Heard mainly at dawn and dusk.

Similar species Chuck-will's-widow is larger, browner, with white necklace and unspotted lower breast; bristles are shorter and straighter. White on male's outer tail feathers is restricted to inner webs. Whip-poor-will (vagrant) is smaller, with black throat, U-shaped necklace.

Habitat and behaviour Rather dense forest, swamp forest, second growth. Breeding March–July. Nest is on leaf-covered ground; lays two greyish-green eggs, spotted with brown. Nocturnal. Feeds on insects caught on wing.

Male has broad white terminal band on tail. Both sexes have dark and light grey mottled plumage, long bristles, white malar stripe and white spots on breast.

Adult.

Range Cuba.

Status Endemic species, common breeding resident. Two subspecies: *A. c. cubanensis* on Cuba and some northern cays, and *A. c. isulaepinorum* on Isle of Pines.

Northern Potoo

Nyctibius jamaicensis 43–46cm

Local name Potú.
Taxonomy Polytypic (5).
Description A tall dark brown or grey bird that perches upright. A complex cryptic pattern with mostly black round spots on chest and scapulars. Large yellow-orange eye, although at night, in torchlight, it can appear red. Tail long and barred. Juvenile duller.
Voice A guttural descending laugh, *kwaah, kwaah, kwaah, kwaah, kwaah, kwaah*.
Similar species None.
Habitat and behaviour Semi-deciduous and evergreen woodland. Nesting unknown on Cuba. Feeds on insects caught on wing.

Adult with white downy chick.

Adult. Note yellow eye, small hooked bill and long tail.

Range Middle America and the West Indies on Cuba, Jamaica and Hispaniola.
Status Very rare resident on mainland Cuba. No breeding data.

Black Swift

Cypseloides niger 15–18cm

Adult. In flight displays all-black plumage and slightly forked tail.

Local name Vencejo Negro.
Taxonomy Polytypic (3).
Description Large, appears blackish overall with pale grey forehead; long wings form an arc in flight; broad short tail, slightly forked in male. Juvenile has white-tipped feathers on underparts. In flight, wingbeats in rapid bursts followed by glides.
Voice A low *chip-chip*.
Similar species Chimney Swift is smaller with shorter square tail, grey-brown plumage and wing shape does not form wide arc; flight is faster with rapid wingbeats. Male Purple or Cuban Martins have broader and more triangular wings.
Habitat and behaviour Over mountain forests. Breeds June–July, but few data. Flight less erratic than other swifts. Feeds on insects caught on wing.

Juvenile has white barring on lower abdomen.

Range Breeds in western North America, Middle America and the West Indies, where *C. n. niger* is common resident on Jamaica and Hispaniola, locally on Cuba and a summer breeder in the rest of the West Indies, excluding the Bahamas and Cayman Islands. Little is known of its migration pattern although the North American subspecies winters in western Amazonia.
Status Uncommon and local breeding resident on Cuba, in Sierra de Guamuhaya, Cienfuegos and Sancti-Spíritus provinces, and in mountains of Holguín, Santiago de Cuba and Guantánamo provinces.

White-collared Swift
Streptoprocne zonaris 22cm

Local name Vencejo de Collar.
Taxonomy Polytypic (9).
Description Large and black with complete white collar, although not always apparent in young birds. Tail forked or squarish. Juvenile is duller, the feathers on body and wing with pale fringes, and less distinct white collar. In flight, long glides.
Voice A loud, piercing *scree-scree-scree*.
Similar species Black Swift is almost entirely very dark brown (appears black in flight). In flight, glides frequently; wingbeats in rapid bursts.
Habitat and behaviour Mountain forest. Very fast flight; often forming flocks of up to 50. Breeds May–July in caves and cliffs near waterfalls; also in hollow Royal Palms, lays two white eggs. Feeds on insects caught on wing.

Adult in flight displays white collar. Note long narrow wings and slightly forked tail.

Adult at nest on a cave ledge.

Range Mexico to northern Argentina; Greater Antilles. *S. z. pallidifrons* is generally considered to be endemic subspecies to the West Indies region.
Status Local but usually common breeding resident on Cuba: in Sierra de Guamuhaya, Cienfuegos and Sancti-Spíritus provinces; and mountains of Holguín, Santiago de Cuba and Guantánamo provinces. Periodically observed in winter on Zapata Peninsula.

Chimney Swift

Chaetura pelagica   13cm

Adult. Note dark upperparts, pale throat and short tail.

Local name Vencejo de Chimenea.
Taxonomy Monotypic.
Description Small swift, with large head, dark grey-brown upperparts, darker rump, pale buffy chin, throat and upper breast; long wings usually only slightly swept back. Tail very short and rounded with feather shafts extending beyond vanes (barely visible). In flight, glides frequently; wingbeats in rapid bursts.
Voice A loud, rapid twittering.
Similar species Black Swift is larger, blackish with longer forked tail.
Habitat and behaviour Over cities and towns, most commonly in coastal areas, observed flying over coastal forest on Zapata Peninsula and Guanahacabibes Peninsula. Hawks insects while on active migration.

Range Breeds in eastern North America, winters in western South America. Rare on passage through the Bahamas, Cuba, Jamaica, Hispaniola, Virgin Islands and Cayman Islands.

Status Near Threatened. Uncommon passage migrant on Cuba and Cayo Coco, September–November and February–May.

Antillean Palm-Swift

Tachornis phoenicobia   11cm

Adult is small with white rump and underparts and black breast-band.

white throat. Both sexes have a dark band across breast. Juvenile similar to female but darker below. In flight, glides frequently; wingbeats in rapid bursts.
Voice Noisy, emitting an almost constant, weak, twittering, *tooee-tooee*.
Similar species Chimney Swift is dark brown with paler throat; tail very short, nearly square with shafts of tail feathers extending beyond vanes.
Habitat and behaviour Grassland and agricultural areas with palms with dense dry fronds. Also in cities with abundant exotic palms with drooping leaves. A colonial breeder from May–July; nest a half-cup among dead palm fronds (*Copernicia macroglossa, C. baileyana, Washingtonia* sp.) with plant fibres and feathers, also under thatched roofs; lays 2–3 white eggs. Flight erratic, bat-like. Feeds on insects caught on wing.

Local name Vencejo de Palma, Vencejito.
Taxonomy Polytypic (2).
Description Very small. Dark brown upperparts with conspicuous white rump and forked tail. Male much darker with white throat; female paler with greyish-

Range Cuba, Jamaica, Hispaniola, Saona and Beata islands, Île-á-Vache.
Status *T.p. iradii* is an endemic subspecies, a very common breeding resident on Cuba and Isle of Pines.

Cuban Emerald
Chlorostilbon ricordii   10cm

Local name Zunzún.
Taxonomy Monotypic.
Description Adult male almost entirely dark green with blue iridescence only rarely visible on breast; white undertail-coverts. Slim. Long, slightly decurved bill with red mandible, white postocular spot and long, forked, black tail. Adult female similar with grey underparts. Juvenile similar to female but back duller green.
Voice Male's song is high-pitched, rapid, rolling series of *slee* notes; also spluttering metallic notes. Female has distinctive flight call, a high-pitched *seeeee*, repeated two to five times.
Similar species Female Bee Hummingbird is much smaller, with much shorter, white-tipped blue tail.
Habitat and behaviour Semi-deciduous and evergreen woodland, rainforest, pine forest, coastal thickets, urban areas. Breeds year-round. Builds a tiny, deep cup-shaped nest of fine fibres coated with lichens and spider webs, in a tree or bush 2–4m above the ground; lays two white eggs. Feeds on nectar, insects, spiders.

Female has entirely grey underparts.

Male is entirely green with bronze and dark blue iridescence, red mandible and long, forked black tail.

Range Bahamas and Cuba.
Status Common breeding resident throughout the Cuban archipelago.

Ruby-throated Hummingbird

Archilochus colubris 9cm

Local name Colibrí.
Taxonomy Monotypic.
Description Adult has iridescent green upperparts, dark wings, white postocular spot, white breast and greenish-white underparts; long, black, thin bill slightly decurved. Tail short and forked, projects slightly beyond wings, legs and feet dark. Male has brilliant iridescent red throat (black unless sunlit), black eye-stripe and forked tail with green central feathers, outer feathers greyish-brown. Female has black eye-line from lores to ear-coverts and whitish throat may have darkish dusky speckles. Tail less forked with white-tipped outer retrices. Immature male resembles female but may have golden cast to upperparts and some red flecks on throat.
Voice Rapid, high *tic* notes.
Similar species Female Bee Hummingbird is smaller with bluish-green back and rounded tail.
Habitat and behaviour Exotic flowering trees in urban and rural gardens, and coastal thickets.

Male has iridescent green crown and upperparts and red throat.

Female has whitish underparts with buffy wash on sides and white tips to outer tail feathers.

Male in flight. Note red on throat, green upperparts and slightly forked tail.

Range Breeds in North America; winters from Mexico to Panama when some cross the 1,000km Gulf of Mexico; rare on passage in the Bahamas, Cuba and Cayman Islands; vagrant elsewhere in the Greater Antilles.

Status Rare passage migrant, mainly to western Cuba, September–November and February–May, with majority of records in spring.

Bee Hummingbird

Mellisuga helenae 6.4cm

Local name Zunzuncito.
Taxonomy Monotypic.
Description The world's smallest bird. Adult male has iridescent deep blue to greenish-blue upperparts, grey underparts. Head, chin and throat iridescent pink, green, golden or red. Tail iridescent blue, short and rounded with black spots on outer retrices. Non-breeding male lacks red on head and throat and so resemble females but with deeper blue back. Adult female is larger, with bluish to green back (more intense blue in breeding), blackish lores and grey underparts; white tips to outer tail feathers. Juvenile similar to female, but duller.
Voice A series of surprisingly loud, very high-pitched, and prolonged whistles and chirps. During breeding season, males vocalise from highest leafless branches.
Similar species Female Ruby-throated Hummingbird is larger, with iridescent green back and slightly forked tail.
Habitat and behaviour Dense forest, semi-deciduous and evergreen woodlands, pine forest, swamp woodland, edge of woodlands, montane serpentine shrub woods. Breeds April–June. Nest a very small neat cup covered with lichens and spider webs; lays two white eggs. Feeds on nectar and insects.

Adult has iridescent blue upperparts, and red crown and throat; red gorget extends to elongated feathers.

Female at cup nest with two young.

Female has metallic sheen on green or bluish upperparts and pale grey underparts.

Range Cuba and Isle of Pines.
Status Near Threatened. Endemic species, rare on mainland Cuba with majority on Guanahacabibes Peninsula, Zapata Peninsula and several eastern mountain ranges. Rare on Isle of Pines.

Male fanning blue tail with black tip on outer feathers.

Adult male.

Adult male.

Cuban Trogon

Priotelus temnurus 27cm

Local name Tocororo, Tocoloro.
Taxonomy Polytypic (2).
Description Iridescent dark green upperparts with violet-blue crown and nape, underparts with white throat and breast, contrasting sharply with vermilion belly. Eye and mandible red. Wings and tail intricately patterned in blue, black, green and white; tips of tail feathers prominently flared.
Voice A rapid, wooden *to-co-lo* or *to-co-lo-ro*, delivered in series, as well as hoarse barking and variety of clucking or chuckling calls.
Similar species None.
Habitat and behaviour Semi-deciduous and evergreen woodland, swamp woodland, tropical karstic woodland. Breeding April–July. Lays 3–4 white eggs in abandoned woodpecker holes, often palms. Feeds on insects, flowers, fruits. Flight undulating.

Adult has iridescent green or blue upperparts. Note wings have white markings, larger on wing-coverts.

Adult. Note conspicuous vermilion abdomen and undertail-coverts.

Adult. Note characteristic, curved lower-back profile.

Adult has red iris and mandible.

Range Cuba and some northern cays.
Status Endemic species. Cuba's national bird. Common and widespread, with two endemic subspecies: *P. t. temnurus* on Cuba and some northern cays (Cayo Guajaba and Cayo Sabinal); and *P. t. vescus* on Isle of Pines.

Adult. Note distinct red lower underparts.

Adult. Note unique tail profile.

Adult.

Cuban Tody
Todus multicolor 10.8cm

Local name Pedorrera, Cartacuba.
Taxonomy Monotypic.
Description Male has bright green upperparts with yellow forecrown; pale grey underparts with brilliant red throat patch; eye bluish; blue patch on side of neck; sides with pink 'powderpuff' effect; undertail-coverts yellow. Bill long, flat with red mandible, short tail. Female slightly duller, and less pink on sides. Juvenile has entirely pale grey underparts and brown eyes.
Voice A hard, rapid chatter *tot-tot-tot* or *trrr*, produced by the wings.
Similar species None.
Habitat and behaviour Present in all manner of woodlands. Breeds April–July. Breeds in burrows excavated into vertical earth banks, sand banks, crab burrows and rotten logs, also in natural cavities in limestone and tree trunks; lays three white eggs. Perches sometimes for long periods, often bobbing head up and down. Flies only short distances, sometimes with a peculiar whirring sound produced by the wings. Feeds on insects, caterpillars, larvae, spiders, even small lizards.

Male displays extensive pink on sides.

Female has less pink on sides.

Adult leaving nest cavity after feeding young.

Range Cuba, Isle of Pines and northern cays.
Status Endemic species. Common and widespread breeding resident on Cuba, Isle of Pines and northern cays.

Adult has red-orange mandible, blue patch on side of neck and yellow undertail-coverts.

Adult male.

Both sexes have vivid green upperparts with red throat.

Belted Kingfisher

Megaceryle alcyon 26–36cm

Local name Martín Pescador.
Taxonomy Monotypic.
Description Adult has blue-grey upperparts, large head with shaggy crest, long heavy blackish bill, broad white collar around throat and nape, underparts white with blue-grey breast-band, and short tail. Male has blue-grey breast-band alone; female also has rufous breast-band and flanks. Juvenile has rusty spotting in the blue breast-band.

Voice Harsh rattle *kek-kek-kek-kek*.
Similar species None.
Habitat and behaviour Wetlands, riparian woodland. Individuals hold winter territories. Wingbeats deep and irregular, showing white wing-patches. Hovers and plunge-dives for fish, also takes crabs, dragonflies; erratic flight with deep wingbeats.

Male has single grey breast-band.

Female has grey upper breast-band and rufous lower breast-band.

Adult in flight displays white patch on outer primaries and white underwing-coverts.

Range Breeds in North America; winters in south-west United States, Middle America to northern South America, and the West Indies.

Status Common winter visitor and passage migrant throughout the Cuban archipelago, but observed in all months.

West Indian Woodpecker

Melanerpes superciliaris 26cm

Local name Carpintero Jabado.
Taxonomy Polytypic (5).
Description Red on head, upperparts and rump finely barred black with cream or gold. Long blackish bill with curved culmen and grey face; tail with long pointed black feathers. Underparts mostly grey, with reddish on abdomen. In flight shows blackish primaries, with white wing-patch at base. Male has brilliant red cap to nape, and red nasal tufts; female has red nape. Juvenile similar to adult but smaller, with less red on head, underparts buffy-grey.
Voice A loud and frequently repeated *kkrraaa* and other loud, chattering calls, resembling those of Fernandina's Flicker.
Similar species Yellow-bellied Sapsucker is smaller, with prominent white patch on inner wing and black-and-white stripes on head.
Habitat and behaviour Woodlands of all kind. Flight undulating. Forages for arthropods, tree-frogs, lizards and many types of fruit by pecking and gleaning in all habitats from mid-levels to near the ground. Breeds February–July, nesting in living and dead tree cavities excavated by pair; 5–6 white eggs. Often nests in the same cavity year after year.

Feeding young at nest cavity.

Female has white forehead and red nape.

Adult male in flight.

Range West Indian endemic with resident subspecies on Cuba, Isle of Pines and Grand Cayman, and two on the Bahamas (probably extirpated on Grand Bahama).
Status There are two endemic subspecies: *M. s. superciliaris* is fairly common breeding resident on Cuba and some northern cays; and *M. s. murceus* breeds on the Isle of Pines and some southern cays.

Yellow-bellied Sapsucker

Sphyrapicus varius 20–23cm

Local name Carpintero de Paso.
Taxonomy Monotypic.
Description Both sexes have red crown, black and white facial stripes, white nape, black breast-band, back and wings heavily barred black and white, white on wing-coverts forming long wide patch, underparts yellowish with blackish chevrons on sides and flanks. Short, straight black bill; outer rectrices black with narrow white border, central tail feathers with broad white and narrow black bars. Male has a red chin and throat. Female has white throat. Juvenile and first-year plumages are olive-brown, head and crown with buff streaks, upperparts black and buff.
Voice A soft *mew*.
Similar species West Indian Woodpecker is larger, lacks stripes on head and black patch behind eye, has red crown and nape and barred wings.
Habitat and behaviour Semi-deciduous and evergreen woodland, swamp woodland, second growth, tropical karstic woodland. Flight undulating. Drills parallel holes around hole of tree and feeds on sap released from these 'wells'.

Adult male has red throat.

Juvenile male, seen here with red on throat beginning to develop.

Adult female has white throat.

Range Breeds in northern North America; winters in North America south to Panama and the West Indies, in the Bahamas, Greater Antilles (rare east of Hispaniola) and Cayman Islands.
Status Common winter visitor and passage migrant throughout the Cuban archipelago, September–May, with majority female and immature.

Cuban Green Woodpecker

Xiphidiopicus percussus 23cm

Male.

Female has black forecrown with white streaks, red hindcrown and nape.

Juvenile has duller plumage compared to adults and black on throat

Local name Carpintero Verde.
Taxonomy Polytypic (2).
Description The only bright green woodpecker on Cuba. Olivaceous-green back; head mostly white, with a black postocular stripe and small red patch on base of throat and upper breast; pale yellow belly streaked with dark brown; bill short; tail narrow with outer feathers barred brown and grey. Male with red from forehead to nape. Female has black forecrown streaked with white, red hindcrown and nape. Juvenile duller; red on head confined to coronal stripe, and blackish throat. Individuals outside mainland Cuba average somewhat smaller and paler; in eastern mountains of Cuba the birds are much larger and more colourful. Some individuals in the northern cays have a bright orange wash on underparts. The population on Isle of Pines is smaller with the amount of red on crown and throat more restricted.
Voice A loud, repeated *taha-taha*; also, a short and harsh, repeated *jorrr* during head-swinging display.
Similar species None.
Habitat and behaviour Occurs in all manner of woodlands. Breeds December–August. Nests in both dead and live tree trunks, termite nests; entrance hole is small and round, 2–12m above ground; lays 3–4 white eggs; all nesting duties shared by both sexes. Drumming is absent in this species. Flight undulating, with noisy wingbeats. The noise of the wings of arriving mate signals the incubating bird to vacate the cavity for an incubation changeover. Forages in all strata; feeds on insects extracted from shallow cracks in bark by inserting bill, as well as into flowers, probably to drink nectar; also, small fruits, frogs, lizards. The most aggressive Cuban woodpecker.

Range Cuba.
Status Endemic species, common and widespread throughout the Cuban archipelago. Several subspecies have been described, but only two are considered valid: *X. p. percussus* from Cuba and *X. p. insulaepinorum* from Isle of Pines.

WOODPECKERS 219

Male has red from forehead to nape.

Adult female.

Male in cavity.

Adult male.

Both sexes display red on throat, yellowish area on central abdomen and streaked and barred underparts.

Northern Flicker
Colaptes auratus 30cm

Local name Carpintero Escapulario.
Taxonomy Polytypic (10).
Description Cinnamon-buff face, sides of neck, chin and throat; grey crown with scarlet triangle on central nape, long pointed black bill; black upper breast crescent, rest of underparts pale yellow to pale cinnamon-buff with bold black spots. Some individuals may have whitish central belly and cinnamon-buff sides. Back and wings brownish-grey barred with black, rump and uppertail-coverts white heavily spotted with black, tail black with shafts of rectrices bright yellow and outer rectrices barred black and buff. In flight underwing and undertail are golden-yellow. Male has broad black moustachial stripe. Female lacks moustache. Juvenile is paler, scarlet on nape is duller and not very well-defined.
Voice *Kee-yar*, descending; *wicka-wicka* repeated quietly; long, loud, fast series *wikwikwikwik*, also drumming on tree bole.
Similar species None.
Habitat and behaviour Semi-deciduous and evergreen woodland, second growth, pine forest, tropical karstic woodland and wetland, from sea level to at least 1,250m. Flight undulating. Forages for arthropods (mainly ants) and wide variety of fruits, seeds and berries in all habitats from mid-level and often on the ground. Breeds January–August in tree cavities, often dead palms; lays 4–6 white eggs, young brooded and fed by both parents; fledges around 28 days.

Male shows pronounced black malar stripe.

Female lacks black malar stripe.

Range The *chrysocaulosus* group is endemic to the Greater Antilles and resident on Cuba, Isle of Pines and cays, and Grand Cayman.
Status Breeding resident, the endemic subspecies *C. a. chrysocaulosus* is uncommon on mainland Cuba, although fairly common in the northern cays.

Fernandina's Flicker
Colaptes fernandinae 34cm

Local name Carpintero Churroso.
Taxonomy Monotypic.
Description The only woodpecker that is barred throughout and lacks any red on head. Male uniformly yellowish-brown barred with black; cinnamon head, with fine blackish streaks from forehead to nape; yellowish on underwings and tail. Male has black moustache-like stripe, lacking in female. Juvenile is duller.
Voice A repeated and increasingly loud *wika-wika-wika-wika-wika* mainly given during the nest changeover; a loud territorial *keer*, and *krrr*, resembling those of the West Indian Woodpecker and a *Kiu-Kiu* territorial call.
Similar species None.
Habitat and behaviour Swamp woodland, semi-deciduous and evergreen woodland with abundant palms. Breeds January–June. Nests mostly in dead palms, and entrance hole quite large; lays 4–5 white eggs. Feeds mostly on ants and insect larvae, usually on the ground, walking or hopping. Flight undulating.

Male has black malar stripe.

Female.

Adult male in tree hole.

Range Widespread but local on Cuba. Not recorded on Isle of Pines or any of the Cuban cays.
Status Vulnerable. Endemic species, rare and declining, mainly found in Sierra del Rosario, Pinar del Río province; Zapata Peninsula, Matanzas province; Najasa, Camagüey province; and northern Holguín province. Threatened by loss of palm habitat.

Both sexes have barred back and tail.

Adult male.

Male in flight. Note barred underparts and tail.

Ivory-billed Woodpecker
Campephilus principalis 48cm

Local name Carpintero Real.
Taxonomy Polytypic (2).
Description Black with white stripes along neck and back and broad white stripe on wings. Tall crest is red in male, black in female. Bill heavy, straight and ivory-coloured. See page 371 for an illustration of this species.

Voice A very distinctive, repeated, high-pitched, nasal *pent*.
Similar species None.
Habitat and behaviour Tall, undisturbed semi-deciduous and pine forests. Breeds April–June. Nest holes in tall, old pines; lays 2–4 white eggs. Feeds on large insect larvae extracted from rotting logs.

Range Formerly south-eastern United States and Cuba. Officially considered extirpated from United States.
Status Critically Endangered endemic subspecies, *C. p. bairdii*. Last century reported from Pan de Guajaibón, Ensenada de Cochinos and Guantánamo mountains. A few individuals may survive in Humboldt National Park; last sightings in 1984–88. On the verge of extinction or possibly already extinct.

Western Wood-Pewee
Contopus sordidulus 16cm

Local name Bobito de Bosque del Oeste
Taxonomy Polytypic (4).
Description Almost identical to Eastern Wood-Pewee, safely distinguished only by voice. Brownish-green upperparts, whitish throat, breast usually has a broad and uninterrupted dark grey band, and whitish belly; distinct whitish wingbars. Bill dusky above, mandible has variable pale orange base with dusky tip, sometimes bill is entirely black. Juvenile has cinnamon-edged feathers on back and wings.
Voice Most common call is a hoarse, descending *pheeer*.
Similar species On average, Eastern Wood-Pewee has narrower breast-band and yellower mandible. Cuban Pewee has bright whitish crescent behind eye and broader, flatter bill.
Habitat and behaviour Urban areas, coastal thickets. Mainly insects caught on the wing.

Range *C. s. saturatus* breeds in western North America, wintering in north-western South America.
Status Very rare passage migrant on Cuba and some cays, September–October and March–April.

Virtually identical to Eastern Wood-Pewee – the two can only be safely separated by their calls.

Eastern Wood-Pewee

Contopus virens   14–16cm

Adult. Note broad wingbars and long primary projection.

Local name Bobito de Bosque.
Taxonomy Monotypic.
Description Adult has dark olive-brown upperparts, broad whitish wingbars and long primary projection, faint eye-ring behind eye, dark bill with pale orange mandible. Dark grey breast almost forms a breast-band, whitish throat, may have yellow wash in centre of belly, and slightly notched tail. Juvenile may have dark bill; cinnamon-edged feathers on back and wings.
Voice Most common call is a descending *pheeer*.
Similar species On average, Western Wood-Pewee has broader breast-band and blacker mandible. Cuban Pewee is paler with bright whitish crescent behind eye and broader bill.
Habitat and behaviour Swamp woodland, semi-deciduous and evergreen woodland, tropical karstic woodland, coastal thickets, urban areas. Takes insects on the wing before returning to same perch.

Adult.

Range Breeds in eastern North America; winters in South America. Migrates mainly through Middle America and more casually (mainly in autumn) through the Bahamas, western Cuba and, rarely, Jamaica and Cayman Islands.

Status Common passage migrant throughout the Cuban archipelago, August–December and March–May, and very rare winter visitor (December).

Cuban Pewee

Contopus caribaeus 16cm

Local name Bobito Chico.
Taxonomy Polytypic (4).
Description Dark olivaceous-grey upperparts, with dark greyish-buff breast and belly. A whitish crescent behind eye is diagnostic. Bill broad at base; mandible yellow-orange. Juvenile has paler mandible and whitish-buff wingbars.
Voice A prolonged and descending whistle, *weeeeooooo*. Also, a thin, repeated *weet*, on one pitch, similar to, but faster than that of La Sagra's Flycatcher.
Similar species Both wood-pewees have narrower bill, dark grey breast, and lack mark behind the eye. Eastern Phoebe is paler below with narrower black bill and unmarked eye.
Habitat and behaviour Semi-deciduous and evergreen woodland, tropical karstic woodland, rainforest, pine forest. Rare at high elevations. Breeds March–June. Nest is cup-shaped, on limb or in fork of tree or shrub, built with fine rootlets and hairs, covered with lichen and moss; lays up to four white eggs, heavily spotted with dark, violet and brown at larger end. Feeds mainly on insects caught on the wing, and frequently returns several times to same perch; quivers tail upon alighting.

Adult. Note the white crescent behind eye and crested head.

Adult. Note broad bill and rather long tail.

Juvenile.

Range West Indian endemic, with endemic subspecies in Bahamas and Cuba.
Status *C. c. caribaeus* is common breeding resident throughout the Cuban archipelago; the endemic subspecies *C. c. morenoi* and *C. c. nerlyi* have more restricted distributions.

Yellow-bellied Flycatcher
Empidonax flaviventris 14cm

Adult has yellow abdomen, throat and eye-ring, white wingbars and pale orange mandible.

Local name Bobito Amarillo.
Taxonomy Monotypic.
Description Greenish-olive upperparts, yellowish underparts, with broad olive wash on breast. Conspicuous whitish or pale yellow eye-ring; short broad-based bill has pale-orange mandible. Well-marked yellowish or whitish wingbars; legs grey. Short and thin tail. Juvenile has buffy wing-bars. Frequently flicks wings and tail.
Voice A whistled *chu-wee*, and a resonant *che-lek*.
Similar species Acadian Flycatcher is paler yellow below, with whitish throat, white lower breast and longer wings. Alder and Willow Flycatchers usually lack distinct eye-ring and have whitish throat, yellowish belly and are brownish-olive above.
Habitat and behaviour Second growth, coastal thickets. Feeds mainly on insects caught on the wing.

Range Breeds in northern North America, wintering in Middle America.
Status Very rare autumn passage migrant, September–October.

Acadian Flycatcher
Empidonax virescens 13cm

Adult. Note yellow eye-ring, whitish wingbars and yellow mandible.

Local name Bobito Verde.
Taxonomy Monotypic.
Description Flat forehead with peaked crown, greenish upperparts, yellowish abdomen and undertail-coverts, with olive wash on sides of breast, whitish throat; narrow, distinct pale yellow eye-ring, broad-based bill with yellow mandible, sharply defined whitish or buffy wingbars and long primary projection. Greyish legs. Juvenile has broad buffy wing-bars.
Voice A rapid ascending whistle *pee-peet-sa*.
Similar species Yellow-bellied Flycatcher is much yellower below, with shorter wings and a more conspicuous eye-ring.
Habitat and behaviour Second growth, coastal thickets. Feeds mainly on insects caught on the wing.

Range Breeds in the eastern United States; winters in Middle America and north-western South America; rare on passage in the northern Bahamas, western Cuba and Cayman Islands.
Status A rare passage migrant, September–November and February–April.

Willow Flycatcher
Empidonax traillii 15cm

Local name Bobito de Traill.
Taxonomy Polytypic (4).
Description Virtually identical to Alder Flycatcher (vagrant), safely differentiated only by voice. Brownish-olive above with conspicuous wingbars. Eye usually without distinct ring. Pale lores. Throat and upper belly white; breast has olivaceous wash, belly a yellow tone. Legs black; mandible yellow. Juvenile has buffy wingbars. Flicks wings and tail.
Voice A buzzy *phitz-bew*; call an emphatic *rit* or *whit*.
Similar species Acadian and Yellow-bellied Flycatchers are greener above with conspicuous yellowish eye-ring.
Habitat and behaviour Urban areas and coastal thickets. Feeds mainly on insects caught on the wing.

> **Range** *E. t. traillii* breeds in southern Canada, and northern, central and north-east United States, wintering in Central America.
> **Status** Very rare passage migrant on Cuba and Isle of Pines, September–October.

Adult. Brownish-olive above, with yellow mandible, white throat, wingbars and inconspicuous eye-ring.

Eastern Phoebe
Sayornis phoebe 18cm

Local name Bobito Americano.
Taxonomy Monotypic.
Description Dark head contrasts with grey upperparts, fine pointed black bill, steep forehead, wingbars inconspicuous, lacks eye-ring. Throat white and belly pale yellowish-grey, dusky side of breast; frequently bobs long tail. Juvenile is browner above, with buffy wingbars.
Voice A repeated, raspy *phoe-be*.
Similar species Cuban Pewee has distinct white crescent immediately behind eye, broad flat bill with yellow mandible. Wood-pewees have well-defined wingbars, darker on breast, and do not wag their tail.
Habitat and behaviour Open semi-deciduous woodland or open coastal thickets. Tail-wagging habit is a good field mark. Feeds on insects from low perch.

> **Range** Breeds in Canada, eastern and central United States, wintering south to Oaxaca, Mexico.
> **Status** Very rare passage migrant, September–December and January–February.

Adult has dark head, black bill and yellowish wash on underparts.

Great Crested Flycatcher

Myiarchus crinitus 20cm

Local name Bobito de Cresta
Taxonomy Monotypic.
Description Olive-brown crested head and back, grey throat and breast contrast with bright yellow belly, bill black with orange-based mandible, whitish wingbars with rufous outer webs to primaries and inner webs of central tail feathers (extensive reddish-brown tail from below). Juvenile is duller.

Voice Varied: a loud whistle, *wheep*; a rolling *prrr-eet*; and a *wheerrup*.
Similar species La Sagra's Flycatcher has greyish-white underparts.
Habitat and behaviour Swamp woodland, semi-deciduous woodland and evergreen woodland. Feeds mostly on insects high in the canopy.

Adult. Note the bright yellow abdomen.

Juvenile has browner upperparts.

Range Breeds in North America; winters in Florida and Middle America to northern South America.
Status Very rare passage migrant on mainland Cuba and Cayo Coco, September–December and February–May.

La Sagra's Flycatcher
Myiarchus sagrae 19–22cm

Local name Bobito Grande.
Taxonomy Polytypic (2).
Description Brownish-olive back; blackish head, regularly crested; grey from chin to upper abdomen, lower abdomen pale yellow; bill entirely black. Wings and tail have dark cinnamon-brown inner webs; wing-coverts tipped whitish forming distinct wingbars. Juvenile duller with less contrasting wing-coverts markings.
Voice Song, a whistled *weeet-zc-weer* or *weeet-ze*. Call, a slightly ascending whistled *weet*, repeated at intervals, sometimes with a squealing tone.
Similar species Great Crested Flycatcher has darker grey breast, brighter yellow belly, and bright rufous on wings and tail.
Habitat and behaviour Evergreen, semi-deciduous woodland and swamp woodland, second growth, tropical karstic woodland, pine forest, riparian woodland and coastal thickets. Approachable, not very active; perches from low to high, with body slanted off vertical. Takes insects by sallying from perch; hover-gleans for arthropods and larvae on leaves; fruit forms large component of diet. Breeds March–July. Builds nest in small tree cavity lined with dry grasses, hairs, rootlets and feathers. Lays four yellowish-white eggs, spotted with brown and violet at larger end; both parents feed young.

Adult has whitish wingbars and dull cinnamon on wings and tail.

Adult. Note the bushy crest and black bill.

Range West Indian endemic. *M. s. sagrae* breeds throughout the Cuban archipelago and Grand Cayman.
Status A common breeding resident throughout the Cuban archipelago.

Eastern Kingbird

Tyrannus tyrannus 23cm

Local name Pitirre Americano.
Taxonomy Monotypic.
Description Head black, upperparts blackish-grey with indistinct wingbars, white underparts with grey wash on breast, square tail black with broad white terminal band, short black bill.
Voice A strident shrill, *kip-kip-kipper*, or *dzee-dzee*.
Similar species Loggerhead Kingbird has buff white terminal tail-band, yellow wash on axillaries and undertail-coverts, and larger bill.
Habitat and behaviour Coastal thickets, second growth, urban areas.

Range Breeds in North America. Winters in South America; on passage mainly through Middle America, the northern Bahamas and western-central Cuba and Cayman Islands.
Status Uncommon passage migrant during autumn migration, July–November; rare in spring, March–April.

Adult has smaller bill than other kingbirds, greyish band on breast and broad white terminal band on tail.

Adult.

Gray Kingbird

Tyrannus dominicensis 22–25cm

Local name Pitirre Abejero.
Taxonomy Polytypic (2).
Description Adult has grey head and upperparts, orange crown patch seldom visible, blackish-grey mask from lores to ear-coverts; dark grey wing-coverts edged pale; whitish underparts with grey wash on breast; long robust, pointed black bill; grey tail is notched. Juvenile has brown-edged feathers on wings and back and lacks crown patch.
Voice A loud and rolling *pit-piteerri-ri-ree*; also a short, fast *pi-tir-re*.
Similar species Eastern and Loggerhead Kingbirds have black head and squared-ended tail. Giant Kingbird is larger with larger bill, blackish head and slightly notched tail.
Habitat and behaviour Open woodland edge, second growth, coastal thicket, swamp woodland, tropical karstic woodland. Takes large flying insects, lizards and fruits. Aggressive and territorial, attacking raptors in flight. Breeds April–July. Builds cup-shaped nest in trees, with twigs, vines and grasses, lined with fine grass and rootlets. Lays 3–4 creamy white or pale pink eggs, speckled and spotted with chestnut-red, pale lilac and grey.

Adult. Note heavy bill and notched tail.

Adult. Note dark lores and ear-coverts.

Range *T. d. dominicensis* breeds on United States Gulf coast, the Bahamas, Turks and Caicos Islands, Greater Antilles, Cayman Islands, Trinidad, Leeward Antilles and northern Venezuela. The western population is a migrant breeder in the Bahamas, Cuba, Cayman Islands and Jamaica, wintering in Panama and northern South America; the eastern population is resident from Hispaniola east to the Virgin Islands.
Status Common summer breeding visitor throughout the Cuban archipelago, mainly February–November.

Giant Kingbird

Tyrannus cubensis 26cm

Adult.

Local name Pitirre Real.
Taxonomy Monotypic.
Description A large kingbird with a massive bill and distinct hooked tip to maxilla. Dark grey upperparts, blackish ear-coverts and white underparts; depending on light the head can look darker; concealed orange crown patch. Wing-coverts and flight feathers margined with white, underwing-coverts pale yellow. Tail brown, slightly notched, sometimes with pale tip. Juvenile lacks orange crown patch.
Voice A loud, burry *tooe-tooe-tooee-tooee-tooee*.
Similar species Loggerhead Kingbird is smaller, with darker head, shorter and thinner bill, and somewhat more distinct white terminal band on blackish squared tail. Gray Kingbird is smaller, with grey head and well-notched tail.
Habitat and behaviour Grassland with scattered *Ceiba* trees, riparian woodland and pine forests. Breeds March to June. Builds cup-shaped nest high in trees, made of twigs, rootlets and grasses without lining. Lays 2–3 creamy-white eggs speckled and spotted with chestnut-red, pale lilac and grey. Feeds on large insects, lizards, small nestling birds.

Adult.

Juvenile.

Range Cuba and Isle of Pines.
Status Endangered. Endemic species; rare but widespread breeding resident on Cuba on Guanahacabibes Peninsula; Sierra de los Órganos, Sierra de Anafe; less rare in Sierra de Najasa, eastern mountain ranges. Also on Isle of Pines. Declining due to habitat loss.

Adult. Note massive bill and slightly notched tail.

Adult.

Adult. Note brown tail and yellowish underwing-coverts.

Loggerhead Kingbird

Tyrannus caudifasciatus 24–26cm

Local name Pitirre Guatíbere.
Taxonomy Polytypic (6).
Description Blackish head with yellow or pale-orange crown patch usually concealed, black bill with bristles at base, grey back, wing-coverts edged whitish and form indistinct wingbars. White underparts with yellow wash on axillaries and undertail-coverts. Square-ended blackish tail has buffy-whitish tip. Juvenile has browner wings and tail with less distinct markings on wing-coverts and lacks crown patch.
Voice A loud, rising, *pit-pit-pit-pit-pit-tirr-ri-ri-reee* and a rolling, almost spluttering chatter.
Similar species Gray Kingbird has grey head, prominent ear-coverts patch and notched tail. Eastern Kingbird has darker upperparts, white underparts, with a pale grey band on the breast, smaller bill and broad white terminal band on square-ended tail. Giant Kingbird is larger, with heavier bill and slightly notched tail.
Habitat and behaviour Semi-deciduous and evergreen woodland, rainforest, riparian woodland, tropical karstic woodland, swamp woodland. Breeds April–July. A cup-shaped nest is usually built high above the ground, with twigs, rootlets and hairs without lining. Lays 2–3 salmon-coloured eggs with reddish-brown and violet markings at the broad end.

Adult. Note black head and square-ended blackish tail with whitish terminal band.

Adult. This species shows a peak on the back of the head – a unique head profile among Cuban birds.

Range West Indian endemic, with resident subspecies in the northern Bahamas, Greater Antilles and Cayman Islands; casual in Florida.
Status *T. c. caudifasciatus* is a common breeding resident throughout the Cuban archipelago.

Scissor-tailed Flycatcher

Tyrannus forficatus 19–38cm including rectrices

Local name Bobito de Cola Ahorquillada.
Taxonomy Monotypic.
Description Pearly-grey head and back; variably pink-tinged back and rump; whitish below with orange-buff abdomen; wing feathers blackish with white feather edges; lower abdomen and undertail-coverts washed pink; pink underwing-coverts; red axillaries; very long, forked black tail with white outer rectrices. Female has shorter tail. Juvenile has greyish-brown upperparts, buffy wash on abdomen and axillaries, and shorter tail.
Voice A harsh *kee-kee*.
Similar species Fork-tailed Flycatcher has black head, dark grey back and entirely white underparts.
Habitat and behaviour Coastal thickets, perches on fences or telegraph wires.

Adult in flight shows blackish inner tail with long outer feathers and reddish axillaries.

Adult male has very long tail.

Juvenile. Note short tail.

Range Breeds in southern United States and northern Mexico, with a small population in the Florida Keys; winters in Florida and Middle America to Panama; rare or vagrant in the West Indies in the Bahamas, Greater Antilles and Cayman Islands.
Status Very rare passage migrant, October–November and January–March.

White-eyed Vireo
Vireo griseus 12.5cm

Adult. Note yellow 'spectacles', white iris and yellow sides.

Local name Vireo de Ojo Blanco.
Taxonomy Polytypic (6).
Description Adult has grey head and nape, greyish-olive back, bright yellow 'spectacles'; white iris diagnostic; dark wings with pronounced whitish or yellowish-white wingbars; whitish underparts with sides washed yellow. Juvenile has greyish-brown eyes, with pale yellow or buffy spectacles.
Voice A distinct and rapid *chick-a-per-weeoo-chick*.
Similar species Thick-billed Vireo has dark iris and yellowish underparts.
Habitat and behaviour Coastal thickets, semi-deciduous and evergreen woodlands. Usually concealed in fairly dense understorey, foraging slowly among leaves for arthropods (mostly insects) and fruits.

Adult.

Adult. Note grey head, wingbars and greenish-greyish back.

Range *V. g. griseus* breeds in south and eastern United States; winters south to Honduras, the Bahamas and western Greater Antilles.

Status Uncommon winter visitor and passage migrant on Cuba, including Isle of Pines and some larger cays, September–April.

Thick-billed Vireo

Vireo crassirostris 14cm

Local name Vireo de las Bahamas
Taxonomy Polytypic (5).
Description Head grey; greyish-olive back; pale-yellow or whitish below, occasionally with olive wash to sides and flanks. Blackish lores; bold yellow 'spectacles' between the eyes and bill, extending to yellow around eyes, or merely to white crescents in some individuals. Two distinct yellowish-white wingbars; brown eyes; thick bill and grey legs. Juvenile resembles adult but head and back feathers are tipped brown, and wingbars poorly defined.
Voice Extremely variable. A sharp *chi-chip-weeeo-chip*. Calls include a buzzy alarm call *shhh*, a low *turrrr*, and a nasal *enk*.

Similar species Immature White-eyed Vireo has yellow on sides alone, smaller head and bill, and less distinct wingbars.
Habitat and behaviour Coastal thickets. Forages for arthropods, slowly examining branches and leaves; also takes fruit. Usually paired and very aggressive towards other vireos. Deep pendant cup nest is suspended from branch by strips of vegetation, lined with leaves; monogamous, female broods two whitish-buff eggs with heavy dark spots, and both parents feed young.

Adult. Note grey head, dark lores and iris, and yellow 'spectacles' and wingbars.

Endemic and local subspecies on northern cays.

Range Bahamas, Cuba, Cayman Islands, Tortue Island (off Hispaniola), and on Providencia and Santa Catalina islands in the western Caribbean Sea.
Status Endemic subspecies *V. c. cubensis* is **Critically Endangered** due to hurricane impact in September 2017 and hotel development.

Cuban Vireo

Vireo gundlachii   13cm

Local name Juan Chiví.
Taxonomy Polytypic (2).
Description Adult has dark olive-grey upperparts,

Adult.

pale yellow underparts; lores and large postocular crescent creamy-white; eye is reddish-brown and appears proportionately large in comparison to other vireos due to narrow area of brown bare skin around the eye; faint wingbars. Juvenile duller, wingbars and crescent behind eye less distinct.
Voice Song, a loud whistling *chuee-chuee*, or *see-see-ri-lo*, or *wee-beee-eeer*, or *whee-tzeeooo*; highly variable. Also a scolding and often repeated *kik* note, a soft rattling call and occasionally a metallic, repeated *poing*. During spring courtship, rapid, repeated *wheet* notes; also a muted *pewt* repeated at intervals.
Similar species The only vireo that appears to have a large eye, lores and postocular crescent creamy-white, and faint wingbars.
Habitat and behaviour Semi-deciduous and evergreen woodland, swamp woodland, coastal thickets, rainforest, riparian woodland, tropical karstic woodland and bushes, mainly at low elevations. Breeds March–August. Cup-shaped nest of grasses, moss, lichen and hair; lays three white, brown-spotted eggs. Feeds on insects, fruits, small lizards.

Adult. Stocky eyes appear large with large postocular crescent. Note faint whitish wing-bars.

Range Cuba, Isle of Pines and offshore cays.
Status Endemic species, common breeding resident throughout the Cuban archipelago. Two widespread subspecies: *V. g. gundlachii* from western Cuba and Isle of Pines, and *V. g. orientalis* from Cayo Coco and eastern Cuba.

Subspecies *V. g. gundlachii* from western and central Cuba has yellowish underparts.

Subspecies *V. g. orientalis* from eastern Cuba has duller underparts.

Adult. Note stout pinkish-grey bill.

Yellow-throated Vireo
Vireo flavifrons 14cm

Adult.

Local name Verdón de Pecho Amarillo.
Taxonomy Monotypic.
Description Adult has crown olive-green, back greenish becoming grey on lower back and rump, well-marked white wingbars, yellow 'spectacles', dark iris; chin, throat and breast yellow, belly and undertail-coverts whitish. Juvenile has upperparts brownish, with buffy-yellow throat and breast.
Voice Harsh alarm call heard infrequently.
Similar species None.
Habitat and behaviour Semi-deciduous and evergreen woodland, swamp woodland, tropical karstic woodland. Solitary or in pairs foraging high in the canopy.

Adult. Note yellow spectacles, throat and breast; bold white wingbars and belly are diagnostic.

Range Breeds in eastern North America; winters from Middle America to northern South America and the West Indies where it is regular in the Bahamas and Greater Antilles and rare in Jamaica and Puerto Rico.

Status Uncommon to locally fairly common winter visitor and passage migrant on Cuba, including Isle of Pines and northern cays, August–May.

Blue-headed Vireo

Vireo solitarius 14cm

Local name Verdón de Cabeza Azul.
Taxonomy Polytypic (2).
Description Olive-green upperparts, distinct wingbars often tinged yellow; bluish-grey head strongly contrasts with bold white 'spectacles'; throat and central underparts white, sides and flanks bright yellow or with yellow-olive wash; short tail with white outer tail feathers.
Voice Scold is a burry, descending *jeeeer-jeer-jeer*.
Similar species White-eyed Vireo has white iris and yellow spectacles. Thick-billed Vireo has portion of spectacles in front of eye strongly yellow.
Habitat and behaviour Semi-deciduous and evergreen woodland. Gleans for insects and fruits.

Adult. Note blue-grey head with white 'spectacles' and white wingbars.

Range *V. s. solitarius* breeds in southern Canada and north-eastern United States; winters in southern United States, Middle America.

Status Very rare winter visitor and passage migrant, November–April.

Warbling Vireo

Vireo gilvus 14cm

Local name Vireo Cantor.
Taxonomy Polytypic (5).
Description Uniform olivaceous-grey upperparts almost concolorous with crown; short, slightly hooked black bill; dark iris; whitish underparts with pale yellowish sides and undertail-coverts. Supercilium whitish with dusky eye-line and pale grey lores giving face a bland look. Wings unmarked. Breeding adult may have variable yellow wash on underparts.
Voice Melodious warbling song; call a harsh nasal *meezh*.
Similar species Philadelphia Vireo has shorter bill, dark lores and brighter and more extensively yellow underparts.
Habitat and behaviour Semi-deciduous and evergreen woodland, swamp woodland, urban areas. Feeds on insects and fruits.

Adult has greyish upperparts, whitish underparts and bright white or whitish supercilium. Wings unmarked.

Range *V. g. gilvus* breeds in southern Canada, central and eastern United States, wintering south to Nicaragua.

Status Very rare passage migrant, September–October and early April.

Philadelphia Vireo

Vireo philadelphicus 13cm

Adult has pale yellow throat and upper breast, white supercilium and black eye-line. Wings unmarked.

Range Breeds in northern North America; winters in Middle America south to Panama.
Status Rare passage migrant; probably a winter visitor, October–November and February.

Local name Vireo de Filadelfia.
Taxonomy Monotypic.
Description Olive-greyish nape to upperparts and grey cap, broad white supercilium, dark eye-line to lores, brown iris, small black bill may have fleshy base; unmarked wings are diagnostic. Variable yellow underparts (fading to whitish on belly), usually brightest yellow on breast; thick grey legs and short tail.
Voice A *shway* note similar to Warbling, Red-eyed and Black-whiskered Vireos.
Similar species Warbling Vireo has unmarked pale grey lores and whitish underparts, with yellow wash confined to sides; crown almost concolorous with olivaceous-grey back. Non-breeding Tennessee Warbler usually has brighter green upperparts, paler underparts, thinner bill and is less deliberate in its movements.
Habitat and behaviour Woodlands, urban areas. Forages at tip of small branches for insects and fruits.

Red-eyed Vireo

Vireo olivaceus 15cm

Adult. Note grey crown with long white supercilium outlined in black.

Range *V. o. olivaceus* breeds in North America; winters in South America and occurs on passage in the West Indies in the Bahamas and western Greater Antilles (Cuba, Hispaniola and Jamaica), and Cayman Islands; vagrant elsewhere.
Status Common passage migrant on Cuba, Isle of Pines and some larger cays, August–November and February–April.

Local name Vireo de Ojo Rojo.
Taxonomy Polytypic (10).
Description Olivaceous-green back contrasts with grey crown and distinct white supercilium outlined by black crown-line and eye-line. Eye red (sometimes brownish). White underparts may have yellowish on sides of breast, flanks and undertail-coverts. No wingbars. Juvenile has brownish wash on back and pale yellow flanks and undertail-coverts.
Voice The classic *shway* of this group of vireos.
Similar species Black-whiskered Vireo has thin blackish malar stripe, and black line above supercilium is less marked.
Habitat and behaviour Semi-deciduous and evergreen woodland, swamp woodland. Stays high on treetops where it moves deliberately through the foliage.

Black-whiskered Vireo

Vireo altiloquus   17cm

Local name Bien-Te-Veo.
Taxonomy Polytypic (6).
Description Olivaceous-green upperparts contrast with grey crown and white or buffy supercilium, outlined by black crown-line and eye-line; narrow black malar stripe ('whisker'). Eye reddish-brown. Unmarked wings, whitish underparts with yellowish on sides of breast, flanks and undertail-coverts. Juvenile has brown eye; bill has pinkish base and lacks or has only faint black whisker.
Voice Frequently repeated three-syllable phrases *tsee-tsee-we* louder and more repetitious and more warbling than Red-eyed Vireo call.
Similar species Red-eyed Vireo has shorter bill and lacks whisker, and supercilium is outlined more conspicuously in black.
Habitat and behaviour Coastal thickets, semi-deciduous and evergreen woodland, riparian woodland, swamp woodland, tropical karstic woodland, mangroves, urban areas. Breeds April–July. Deep cup-shaped nest suspended below a branch fork, pendant, anchored by strips of fibre; monogamous, female incubates 2–3 whitish eggs spotted with black and purplish-brown speckles; both parents feed young. Feeds on insects and fruits from mid-height to canopy.

Adult.

Adult. Note grey cap with white eyebrow, outlined black, and distinct black malar stripe.

Range *V. a. barbatulus* breeds in southern Florida, the Bahamas, Cuba, Little Cayman and Cayman Brac; winters to Amazonia.
Status Very common summer breeding visitor, February–November; most arrive around March.

Cuban Palm Crow

Corvus minutus   43cm

Adult in flight. Note heavy bill and broad rounded wings.

Local name Cao Pinalero, Cao Ronco.
Taxonomy Monotypic.
Description Entirely black with very slight metallic sheen. Smaller bill than Cuban Crow, bristles (un-angled) always covering nostrils; short wings. Juvenile is duller.
Voice A nasal, complaining *craaah* note, sometimes repeated in pairs, vaguely reminiscent of Fish Crow (*C. ossifragus*).
Similar species Mainly differentiated by voice and behaviour. Cuban Crow is slightly larger; bristles covering nostrils are at an angle.
Habitat and behaviour Open country with abundant Royal Palms; forests near farms. Breeds March–July; nests on the fronds of Royal Palms; nest very similar to that of Cuban Crow; average clutch size of four eggs, pale blue with well-defined markings slightly concentrated at larger end. Omnivorous, most often seen on the ground. May form small flocks. Typically flicks tail upwards on alighting.

Adult. Note smaller bill than Cuban Crow; bristles cover nostrils.

Range Cuba.
Status Near Threatened. Endemic species. Restricted breeding resident on Cuba to Sierra de Najasa, Tayabito, Miguel, and El Jardín, Camagüey province; La Pedrera, Paloma, near Trinidad, Sancti-Spíritus province; locally common in Camagüey province (Najasa municipality).

Adult.

Adult.

Adult.

Cuban Crow

Corvus nasicus 46cm

Local name Cao Montero.
Taxonomy Monotypic.
Description Entirely black with a slight metallic sheen. Black bill longer than Cuban Palm Crow. Breast with slight grey wash (only visible at close range). Bristles at an upward angle regularly expose the nostrils. Juvenile is duller with some browner feathers.
Voice Very noisy, producing a variety of rather high-pitched, warbled and often parrot-like calls, sometimes rhythmically repeated, *kweaa*, *kraak*, or *kaaaa*. Common call is *gawwaaak-gawow*.
Similar species Cuban Palm Crow is slightly smaller, with shorter wings and bill. Feathers at base of maxilla always cover nostrils; regularly seen on ground. Often best differentiated by voice.
Habitat and behaviour Rainforest, semi-deciduous and evergreen woodlands, clearings near woodlands, palm groves, borders of swamp woodlands. Breeds March–July. Builds nest on large bromeliads or among palm fronds, constructed of twigs, dry grasses and feathers; lays up to four greenish eggs, spotted with brown and lilac. Omnivorous. Usually occurs alone or in pairs, but larger groups may gather, especially near communal roosts. Noisy. Rarely seen on ground, an important field distinction.

Adult in flight.

Adult. More robust bill than Cuban Palm Crow; bristles do not normally cover nostrils.

Range Cuba and southern Bahamas (Providenciales and Caicos).
Status Common breeding resident on Cuba, Isle of Pines and some larger northern cays. Especially common on Guanahacabibes Peninsula, Zapata Peninsula, Sierra de Najasa and Sagua–Baracoa mountain range.

Purple Martin

Progne subis 20cm

Local name Golondrina Azul.
Taxonomy Polytypic (3).
Description Male is entirely very dark metallic purplish-blue. Female duller with very dark iridescent purplish-blue upperparts, grey forecrown and hindneck; throat, upper breast and sides scaly greyish-brown, fading to white on streaked belly. Juvenile has grey-brown upperparts and breast, and greyish-white underparts. In flight fast flapping alternates with short glides; tail forked.
Voice A rich, repeated *tew* and melodious warbled notes with a throaty, plucking quality.
Similar species Male Cuban Martin is virtually indistinguishable, but female may lack streaks or show faint streaks on belly and darker brown on breast and sides.
Habitat and behaviour Wetlands, riparian woodland, urban and rural areas. Takes insects on the wing.

Adult male (right) is glossy blue-black overall.

Female has scaling effect on greyish-brown breast, pale nuchal collar and streaked abdomen.

In flight displays long wings with head noticeably extending beyond the leading edge of the wings; forked tail.

Range *P. s. subis* breeds in southern Canada, central and eastern United States; winters in South America to northern Argentina.

Status Common passage migrant on Cuba, Isle of Pines and some cays, August–November and January–March.

Cuban Martin

Progne cryptoleuca 20cm

Local name Golondrina Azul Cubana.
Taxonomy Monotypic.
Description Male resembles Purple Martin except white feathers on lower abdomen are visible at close range. Female has brown throat, breast and sides; belly unstreaked or with faint streaks. Juvenile similar to female.
Voice Very similar to that of Purple Martin but more metallic.

Similar species Female Purple Martin has more densely streaked belly and lacks the dark brown on sides and breast; greyish on hindneck.
Habitat and behaviour Open country, urban areas, riparian woodland, wetlands. Breeds April–August. Colonial nester, but scattered pairs may breed alone. Lays 3–5 white eggs in abandoned woodpecker holes or natural tree cavities; also in old buildings. Catches insects on the wing.

Male has purple-glossed plumage with white flecks on abdomen.

Female.

Male is almost completely purple with some white markings on lower belly.

Female in flight shows brown throat, breast and sides, which contrast with white underparts.

Range Cuba archipelago. Winter range unknown, presumably South America.
Status Common summer breeding visitor on Cuba, Isle of Pines, Cayo Coco and Cayo Romano, January–November.

Tree Swallow

Tachycineta bicolor 14cm

Local name Golondrina de Árboles.
Taxonomy Monotypic.
Description Male has dark metallic blue-green upperparts, brown wings, white underparts and slightly forked tail; in flight shows white extending onto lower back. Female resembles male, usually slightly duller; both sexes greener in autumn. Juvenile has greyish-brown upperparts and white underparts.

Voice A twittering *klweet*.
Similar species Bahama Swallow has violet-blue gloss on wings, rump and tail, which is deeply forked; white wing-linings.
Habitat and behaviour Wetlands and grassland. On migration, hawks over open ground and shores with other swallows. Feeds on insects caught on the wing.

Adult in flight. Note white underparts, brown wings and forked tail.

Adult. Note blue-green upperparts and entirely white underparts.

Adults in breeding plumage have blue-green upperparts and slightly forked tail.

Range Breeds in North America; winters from southern North America south to South America, and the West Indies in the northern Bahamas, Cuba, Hispaniola, Jamaica and Cayman Islands.

Status Common winter visitor and passage migrant throughout the Cuban archipelago, September–June.

Bahama Swallow

Tachycineta cyaneoviridis   15cm

Local name Golondrina de Bahamas.
Taxonomy Monotypic.

Adult male in flight. Note deeply-forked violet-blue tail and wings. White underparts and wing-lining are diagnostic.

Description Male green on head and back, with violet-blue gloss on wings, rump and tail; white underparts and underwing-coverts. Tail deeply forked. Female duller, with shorter tail. Juvenile even duller brown with shorter tail.
Voice A low, continuously repeated *chi-weet*.
Similar species Tree Swallows wintering on Cuba have mostly brown or blackish-brown wings, and tail only slightly forked.
Habitat and behaviour Open fields with low vegetation. Feeds in flocks, with very rapid flight hawking for insects.

Range Breeds in northern Bahamas, wintering throughout the Bahamas and eastern Cuba.
Status Very rare winter visitor, September–April.

Northern Rough-winged Swallow

Stelgidopteryx serripennis   12.5–14.0cm

Local name Golondrina Parda.
Taxonomy Polytypic (6).
Description Entirely warm brown above with paler rump, brownish chin, throat, breast and sides; rest of underparts whitish. Broad wings. Tail slightly forked. Juvenile has cinnamon on upperwing-coverts.
Voice Call a series of soft *brrrt* notes repeated.
Similar species Bank Swallow has grey-brown upperparts and broad breast-band on white underparts.
Habitat and behaviour Wetlands, grasslands. Hawks for flying insects over open ground, usually with Barn Swallows.

Adult. Note pale brownish throat, breast and sides.

Adult in flight. Brown above, with brownish throat and breast, white belly and forked tail.

Range *S. s. serripennis* breeds in south-east Alaska, Canada, United States, wintering in Florida south to Panama; rare on passage in the Bahamas, Turks and Caicos Islands, and most of Greater Antilles and Cayman Islands.
Status Uncommon passage migrant on Cuba and Isle of Pines, most often seen during spring migration; rare winter visitor, July–May.

Bank Swallow (Sand Martin)

Riparia riparia 13cm

Local name Golondrina de Farallones.
Taxonomy Polytypic.
Description Greyish-brown upperparts, white underparts with a wide brown breast-band; tail slightly forked. Juvenile has buffy wingbars.
Voice A buzzy, incessantly repeated *bijzzz*.
Similar species Northern Rough-winged Swallow is more warmly toned above, lacks breast-band, and flight more graceful with deeper and slower strokes.
Habitat and behaviour Coastal areas, grasslands. On migration, some individuals occur among large flocks of Barn Swallows, hawking for insects.

Range *R. r. riparia* breeds in North America, wintering from Panama to South America.
Status Very rare passage migrant on Cuba and Isle of Pines, August–October and January–June.

Adult. Note brown breast-band.

Cliff Swallow

Petrochelidon pyrrhonota 13–15cm

Local name Golondrina de Farallón.
Taxonomy Polytypic (4).
Description Very dark glossy blue upperparts with white streaks, white forehead (south-western subspecies has rusty forehead), sides of neck dark cinnamon with greyish collar and buffy-orange rump; chin and throat cinnamon to blackish, abdomen whitish. Tail square. Juvenile has brownish-grey upperparts (except rump); tertials have buffy-rufous edging.
Voice A short, melodious *chur*, as well as peculiar squeaky notes.
Similar species Cave Swallow has rufous forehead and throat.
Habitat and behaviour Coastal, wetlands, grasslands.

Range *P. p. pyrrhonota* breeds in North America and Mexico; winters in South America; on passage in the West Indies where rare in the Bahamas, Cuba, Virgin Islands and Cayman Islands; vagrant in Jamaica and Hispaniola or, more likely, overlooked among breeding Cave Swallows.
Status Rare passage migrant, August–November and March–June.

Adult in flight. Note white forehead, rufous face and blackish-cinnamon throat.

Cave Swallow

Petrochelidon fulva   12cm

Adult at nest. Note rufous forehead, throat and collar.

Local name Golondrina de Cueva.
Taxonomy Polytypic (6).
Description Adult has dark blue crown and white streaking on dark blue back; rufous forehead, cheeks, throat, wide collar and rump; buffy breast, whitish belly; and square tail.
Similar species Cliff Swallow has white forehead and dark throat (cinnamon to black).
Voice A melodious *chur* and chattering calls.
Habitat and behaviour Wetlands, coasts and urban areas with Barn Swallows, usually seen on the wing. Breeds March–August. Nest is built mainly of mud, on cliffs, in caves, under bridges or eaves; lays 3–4 white eggs, heavily spotted with brown.

Adult in flight. Note short, square-ended tail.

Range *P. f. cavicola* breeds on Cuba and Isle of Pines, expanding to Florida in the late 1980s; wintering grounds unknown. Populations of this species complex breed in south-central United States, Mexico, Ecuador, Peru and Greater Antilles. Populations are mostly migratory on Cuba (breeds in summer) and mostly sedentary on Jamaica, Hispaniola and Puerto Rico. Vagrant in the Lesser Antilles.
Status Common summer breeding visitor and passage migrant on Cuba, Isle of Pines, and northern cays, December–November.

Barn Swallow

Hirundo rustica 15–19cm

Local name Golondrina de Cola de Tijera.
Taxonomy Polytypic (8).
Description Adult has rufous forehead and throat; crown and upperparts steel-blue more iridescent in male, blackish wings, tail deeply forked with elongated outer tail feathers and white tail-spots. Underparts pale to rich cinnamon-rufous. Female is less glossy than male and has shorter tail-streamers. Juvenile has brownish wash on blue upperparts, pale cinnamon throat, rest of underparts whitish, and short tail. In flight shows deeply-forked tail with white subterminal band.
Voice Twittering song, and *wit-wit* call while hawking.
Similar species No other adult swallow has long tail-streamers and tail-spots.
Habitat and behaviour Usually in flight over urban areas and grasslands, wetlands and coasts.

Adult. Note dark throat and long tail-streamers.

Adult in flight shows deeply-forked long tail.

Adult.

Range *H. r. erythrogaster* breeds in North America, Mexico and Argentina; winters in Middle and South America, Puerto Rico and the Lesser Antilles; on passage throughout the West Indies. Cosmopolitan.

Status A common passage migrant throughout the Cuban archipelago, August–November and February–June.

Zapata Wren

Ferminia cerverai 16cm

Local name Ferminia.
Taxonomy Monotypic.
Description Upperparts brown, finely but densely barred with blackish. Head streaked on top, distinct grey lores, and may show faint supercilium. Bill long and slight decurved, darker on maxilla, paler on lower; brown iris; tail long, densely barred on top and broadly barred with whitish underneath, usually held pointing down when bird is perched. Underparts greyish with whitish throat, brown sides and densely barred flanks; some individuals may have barred sides with dark speckles on breast (may form very faint streaks with tiny spots on neck-side, only seen at close range), barred undertail-coverts, grey legs. A poor flier with very short, rounded wings.

Voice Male song is a pleasant, loud, canary-like warbling beginning with one to three sweet whistled introductory notes, followed by a complex series of grating rattles and buzzes, often repeated in pairs or triplets. Call notes, usually given by the female, include a guttural *kraok* and often repeated metallic *tik* notes not unlike two stones clicked together.
Similar species None.
Habitat and behaviour Marshes with extensive fields of sawgrass and patches of shrubs seasonally flooded to a depth of about 0.5m. Breeds March–May. Nest is globular with a side entrance; known to lay at least two eggs, but few nests have been found. Skulks and is often difficult to observe. Feeds on insects, spiders, small fruits, molluscs, lizards.

Adult. Note greyish underparts and barred flanks.

Zapata Wren is brown overall with streaked head; back and long tail are barred black and brown.

Range Cuba.
Status Endangered. Endemic species, restricted to Zapata Peninsula where it is known from the western section of the marsh.

Adult.

Adult.

Ruby-crowned Kinglet
Regulus calendula 11cm

Adult. Note white crescents before and behind eye. One of the two white wingbars is more prominent than the other.

Local name Reyezuelo.
Taxonomy Polytypic (3).
Description Very small and plump, appearing almost neckless. Olivaceous-grey upperparts and pale yellowish-grey underparts; prominent wide arcs before and behind eye, and two white wingbars (one almost indistinct; the other bordered black). Male has concealed scarlet crown patch. Bill very short and thin.
Voice A fast, sharp *ji-dit*.
Similar species Some vireos and warblers have basically similar colour patterns but are larger and proportionately slimmer, with longer and heavier bills.
Habitat and behaviour Coastal thickets, semi-deciduous woodlands.

Adult male displaying scarlet crown patch.

Range *R. c. calendula* breeds in northern and western North America, wintering south to Mexico and Guatemala.

Status Very rare winter visitor, October–late February.

Blue-gray Gnatcatcher

Polioptila caerulea 11cm

Local name Rabuita.
Taxonomy Polytypic (8).
Description Non-breeding adult has bluish-grey upperparts, wings unmarked, whitish-grey underparts. Conspicuous white eye-ring, fine blackish bill with mostly pale mandible and long black tail with white outer tail feathers (shows as white undertail when perched). Male breeding has U-shaped black line (variable in length), usually from forehead to over the eye, absent in female. Juvenile has brownish wash on upperparts.
Voice Common call note is a mewing, buzzy *zee, zee-zee* or *speeee*. Song is an insect like, thin, high warble.
Similar species Cuban Gnatcatcher has black ear-coverts crescent, and louder and more varied vocalisations.
Habitat and behaviour Semi-deciduous and evergreen woodland, swamp woodland, tropical karstic woodland, riparian woodland. Sings from February on Cuba. Song is an insect-like, thin, high warble. Feeds on small insects. Frequently raises tail while feeding high in the canopy.

Breeding male has black eyebrow.

Non-breeding male and female are paler than breeding male, and eyebrow is absent.

Range *P. c. caerulea* breeds in Canada, central and eastern United States, wintering south to El Salvador, the Bahamas, Cuba and Cayman Islands.

Status Common winter visitor on Cuba, Isle of Pines and some larger cays, August–May.

Cuban Gnatcatcher

Polioptila lembeyei   11cm

Adult.

Local name Sinsontillo.
Taxonomy Monotypic.
Description Difficult to separate sexes in the field. Bluish-grey above with narrow black crescent in ear-coverts area usually bolder in males, distinct white eye-ring, bill black and thin. Plain wings with distinct blackish alula; pale grey underparts, darker on sides. Slim and restless. Frequently cocked tail is long and black, with white in outer feathers, more extensive in outermost feather (underside view shows folded tail mostly white with distinct black markings). Juvenile is paler, with an olive wash on back, ear-coverts crescent barely evident.
Voice Song, a rather loud and long rambling series of warbles, whistles, and chattering notes *psss-psss*, *tiizzzzz-tzi-tzii*, etc., disorganised but sustained; also, an incessantly repeated *pip* or *pyip*.
Similar species Wintering Blue-gray Gnatcatchers generally lack black facial markings, although late-winter males show black supercilium.
Habitat and behaviour Coastal thickets, scrub vegetation. Breeds March–July. Nest is cup-shaped, larger but otherwise not unlike those of hummingbirds. Lays 3–5 white eggs spotted with brown. Forages low in vegetation for insects, spiders.

Adult female has smaller crescent than male.

Adult.

Range Cuba.
Status Endemic species. Common, but local, mostly restricted to coastal localities in eastern section of the mainland, but also found inland: Punta Maisí, Baitiquirí, Cabo Cruz and central Cuba, around Casilda and Trinidad and Sierra de Cubitas. Also on the northern cays: Cayo Coco, Cayo Romano and Cayo Paredón Grande, where the population declined dramatically after the September 2017 hurricane.

Adult male has bold black ear-coverts crescent.

Cuban Solitaire

Myadestes elisabeth 19cm

Local name Ruiseñor.
Taxonomy Polytypic (2).
Description Olivaceous-brown upperparts (may show a rufous wash on face and on back) with distinct creamy-white eye-ring broader behind it; darkish bill with pale base and long bristles; wing unmarked with distinct rufous edges on secondaries, pale grey underparts with conspicuous brown malar stripe. Long tail with white-edged outer feathers, usually held straight down below body. Juvenile has more olive upperparts, with brown flecks on underparts; tertials have buff tips.
Voice Song is a series of sustained, metallic, flute-like, almost harsh *zheeee* notes delivered very deliberately on varied pitches, interspersed with briefer warbled or trilled phrases. Cuba's best songster, more often heard than seen when perched.

Adult has creamy eye-ring, black whiskers and pale legs.

Adult.

Adult.

Similar species La Sagra's Flycatcher has cinnamon-brown tail, lacks malar stripe and commonly holds tail in line with back.
Habitat and behaviour Semi-deciduous and evergreen woodland, tropical karstic woodland, cloud forest, rainforest and pine forests near limestone cliffs. Breeds May–July. Nest is a cup constructed of rootlets, hair and lichens, built in crevices in limestone cliffs as well as in tree cavities; cavity entrance is usually shielded by bromeliads. Lays three pale green, brown-spotted eggs. Feeds while hovering in trees for insects, fruits, seeds.

Range Cuba.
Status Endemic species. Common, but restricted to eastern and western mountain ranges of Cuba. Two subspecies described: *M. e. elisabeth* from Cuba, and *M. e. retrusus* from Isle of Pines, possibly now extinct.

Adult.

Adult.

Juvenile.

Veery

Catharus fuscescens 16–18cm

Adult. Note reddish-brown upperparts and fine spots on buffy breast.

Local name Tordo Colorado.
Taxonomy Polytypic (4).
Description Medium-sized thrush. Adult upperparts vary from bright tawny to reddish-brown, pale indistinct grey eye-ring and face, white throat, may have indistinct malar stripe, fine reddish to brown spots on breast washed evenly with warm buff, sides greyish and rest of underparts whitish; bill dark with pale mandible, legs pale. First-year plumage similar but pale edges to wing-coverts.
Voice Silent.
Similar species Less spotting on breast than other migrant thrushes, and upperparts reddish-brown not greyish-brown.
Habitat and behaviour Semi-deciduous and evergreen woodland with dense understorey; forages in leaf litter for insects.

Range Breeds in North America; winters in Brazil. On passage, through Middle America and the West Indies.
Status *C. f. fuscescens* is a rare passage migrant on Cuba and larger northern cays in autumn, and on Cayo Anclitas, off south coast, August–October and April–May; three winter records. *C. f. salicicola* is a vagrant.

Bicknell's Thrush

Catharus bicknelli 19cm

Local name Tordo de Bicknell.
Taxonomy Monotypic.
Description Very similar to Gray-cheeked Thrush, but browner above, with chestnut tail. Bill blackish

Adult. Note extensive bright-yellow orange mandible and chestnut tail.

above with extensive bright yellow-orange base to mandible, and greyish-brown lores and ear-coverts. Indistinct pale eye-ring, regularly with a broad crescent behind the eye. Under most circumstances, the two species cannot be safely differentiated in the field, only by voice.
Voice Nasal *pheu*.
Similar species Gray-cheeked Thrush is more olive above, lacks chestnut tail and has pale legs. Swainson's Thrush has conspicuously buffy eye-ring, lores and breast.
Habitat and behaviour Dense scrub within cloud forest, mixed evergreen and semi-deciduous woodland. Feeds on insects and fruits.

Range Breeds in north-eastern North America; winters mainly on Hispaniola.
Status Very rare winter visitor and passage migrant on Cuba, observed in Botanical Garden of Havana, and in small numbers on Pico Turquino, Pico Suecia, Pico Cuba and Pico Botella (1,426–1,960m above sea level), October–April.

Gray-cheeked Thrush
Catharus minimus 16–20cm

Local name Tordo de Mejillas Grises.
Taxonomy Polytypic (2).
Description Grey-brown upperparts, greyish face with faint white streaking on ear-coverts, indistinct pale eye-ring, regularly with crescent only behind eye, dark malar stripe, whitish throat, heavy blackish-brown triangular spots on entire whitish breast (some with buffy wash) becoming blurred on whitish belly, sides olive-grey, and legs pale. First-year plumage similar but wing-coverts edged pale.
Voice Silent.
Similar species Swainson's Thrush has distinct buffy 'spectacles'; spots on buffy breast are brownish. Bicknell's is warmer brown above, particularly on tail and has more yellow on lower bill.
Habitat and behaviour Woodland with dense understorey, observed hopping in leaf litter foraging for insects; also takes fruit.

Adult. Note grey face, indistinct pale eye-ring, regularly showing crescent behind eye, and heavy blackish-brown triangular spots on whitish breast.

Adult.

Range Breeds in northern North America and extreme north-eastern Siberia; winters in northern South America. Occurs on passage through Middle America, rarely in western Greater Antilles and Cayman Islands.

Status Rare passage migrant on Cuba (*C. m. aliciae* and *C. m. minimus*) and the cays, September–November and March–May.

Swainson's Thrush
Catharus ustulatus 17.5cm

Adult.

Local name Tordo de Espalda Olivada.
Taxonomy Polytypic (6).
Description Olive-brown upperparts, buffy cheeks and throat, distinct creamy-buff 'spectacles', triangular brown spots on breast becoming blurred reddish-brown wash on sides and flanks, bill dark with pale base to mandible, pale legs. First-year plumage similar but wing-coverts edged buff.
Voice Silent.
Similar species Gray-cheeked and Bicknell's Thrushes have inconspicuous whitish eye-ring and lores; breast usually white sometimes buffy; greyish upperparts and grey cheeks.
Habitat and behaviour Forages on the ground in semi-deciduous woodland and forest edges for insects and takes fruits.

Adult. Note distinct creamy-buff 'spectacles'.

Range Breeds in North America; winters from Mexico to northern Argentina, and on passage in the Bahamas, Cuba, Jamaica and Cayman Islands.

Status *C. u. swainsoni* is uncommon passage migrant on Cuba, Isle of Pines, and several northern cays, September–November and March–May.

Wood Thrush

Hylocichla mustelina   20cm

Local name Tordo Pecoso.
Taxonomy Monotypic.
Description Large thrush, white below with bold blackish spots on breast, sides and flanks. Rufous-brown crown, nape and neck, darker rufous-brown on back and tail, pale lores and pronounced eye-ring, streaked ear-coverts, legs pale flesh. Juvenile has brownish upperparts streaked with buff and spotted underparts.
Voice Snappy *wit-wit-wit-wit* and lower *tut-tut-tut-tut*.
Similar species Swainson's, Gray-cheeked and Bicknell's Thrushes are smaller, lack bright rusty tones, and are less extensively and boldly marked below.
Habitat and behaviour Semi-deciduous and evergreen woodland, and in trees with dense understorey. Feeds on fruits and insects.

Adult has white eye-ring and large black spots on breast and sides.

Range Breeds in North America; winters in Middle America to Panama; rare on passage in the Greater Antilles and Cayman Islands.

Status Rare passage migrant on Cuba, Cayo Coco and Cayo Guajaba, September–November and February–April. A few winter records.

American Robin

Turdus migratorius   25cm

Local name Tordo Migratorio.
Taxonomy Polytypic (7).
Description Large bird. Dark grey above with blackish head, bold broken white eye-ring, and commonly with a supraloral spot, yellow bill that may be dark on top with dark tip, white-tipped blackish tail. Underparts reddish-brown; lower abdomen white; black-streaked white throat; white vent and undertail-coverts. Female much duller overall with white areas buffer. Juvenile has buff spots on back.
Similar species None.
Voice Staccato *tut-tut-tut;* in flight, a thin *see-lip*.
Habitat and behaviour Open country. Feeds on insects and fruits.

Adult. Note blackish head and tail; reddish-brown underparts are diagnostic of this species.

Range North America and Mexico, wintering south to Guatemala and rarely to the northern Bahamas.
Status Very rare passage migrant to Cuba (*T. m. migratorius*), mostly occurring in the western half, and on Cayo Coco and Cayo Levisa in the northern cays, September–December and March–April.

Red-legged Thrush
Turdus plumbeus 25–28cm

Eastern Cuban subspecies has grey lower abdomen.

Local name Zorzal Real.
Taxonomy Polytypic (6).
Description Large bluish slate-grey thrush with orange-red legs. Adult has dark slate upperparts, wings black edged with grey, blackish-grey lores, red-orange eye-ring; bill dark reddish-orange, becoming all dark (season dependent); white chin and black throat. Lower abdomen grey in eastern populations, reddish-brown in birds from Holguín province, central and western Cuba and Isle of Pines. Undertail-coverts white, long black tail showing as white tips to outer tail feathers in flight. Juvenile duller, wing-coverts mottled buffy, underparts greyish, throat and upper breast spotted blackish, bill blackish.
Voice A series of creaks and whistles, commonly uttered in pairs with a distinct pause between each note *chirruit*-(pause)-*chirruit* or *pert*-(pause)-*squeeer* or *squit*-(pause)-*seeer*. When alarmed and taking wing, *wet-wet*; also a mewing note reminiscent of Gray Catbird.
Similar species None.
Habitat and behaviour Semi-deciduous and evergreen woodland, pine forests, tropical karstic woodland, riparian woodland, rainforest, cloud forest, coastal thickets, urban areas. Breeds March–November; builds bulky nest of grasses and dry leaves, lined with hairs, feathers and plant fibres; lays 3–5 greenish-white eggs spotted with brown. Feeds mostly on the ground on insects, also fruits on trees.

Western and central Cuban subspecies has rusty-orange patch on lower abdomen and vent.

Adult. Note mostly grey plumage, with white-tipped outer tail feathers, and red-orange eye-ring and legs.

Range West Indian endemic with subspecies in the northern Bahamas, Cuba, Cayman Brac, Hispaniola, Puerto Rico, and Dominica in the Lesser Antilles.
Status Endemic subspecies *T. p. rubripes* is common breeding resident in western and central Cuba, Isle of Pines and throughout northern cays; *T. p. schistaceus* is common throughout eastern Cuba.

Gray Catbird

Dumetella carolinensis 23cm

Local name Zorzal Gato.
Taxonomy Monotypic.
Description Dark slate-grey upperparts, paler underparts, orange-brown undertail-coverts; black forehead, crown and tail, short wings, slate-grey bill and legs. Juvenile is paler below.
Voice Cat-like mewing *mea*, and short low-pitched *churr*.
Similar species None.
Habitat and behaviour Usually concealed in understorey. Solitary and shy, heard more often than seen, except during migration when small flocks are usual. Flicks tail or holds it erect with wings drooped.

Adult. Note wing-flicking display.

Adult in flight shows overall grey plumage, black crown and reddish undertail-coverts.

Range Breeds across southern Canada, central and eastern United States and Bermuda; winters in extreme south and south-eastern United States, the Caribbean and Atlantic coastal areas of Middle America south to Colombia.
Status Very common winter visitor and passage migrant throughout the Cuban archipelago, August–May.

Northern Mockingbird

Mimus polyglottos   23–28cm

Local name Sinsonte.
Taxonomy Polytypic (2).
Description Dull grey above and whitish below, with large white wing-patch and long tail with white margins. Bill and legs blackish; eye yellow. Juvenile is streaked below, with brownish eye, and buffy wing-coverts.

Voice Varied, loud, and melodious phrases; many are imitations of other birds' calls. Common call is a harsh *tchek*. Male's delightful song in breeding season may be sustained for tens of minutes and is often delivered at night.
Similar species Bahama Mockingbird, restricted to northern cays, is larger (lacks white wing-patch), has black malar streaks on sides and back and faintly on upper breast, and broad long white-tipped tail.
Habitat and behaviour Open disturbed habitats, second growth, urban and littoral areas, coastal thickets. Feeds mainly on insects and fruits. While foraging, may hold tail erect and wings drooped; sometimes raises wings, exposing white patches. Highly territorial. Breeds February–August; nest is bulky cup of coarse dead twigs, weed stems, decayed leaves, rags, string, and cotton, lined with fine grasses, rootlets and hair, placed in dense bushes or low in trees. Lays three pale bluish-green eggs, spotted and blotched with brown.

Adult in characteristic pose with drooped wings and raised tail.

Juvenile. Note wing-flashing displays while foraging.

Range Breeds in North America, Mexico and West Indies in the Bahamas, Greater Antilles, Cayman Islands and Virgin Islands.
Status *M. p. orpheus* is a common breeding resident throughout the Cuban archipelago.

Bahama Mockingbird
Mimus gundlachii   28cm

Local name Sinsonte Prieto, Sinsonte de Bahamas.
Taxonomy Polytypic (2).
Description Brownish-grey above, paler below; back, upper breast and sides sparingly streaked with dark brown; blackish malar stripe, wings lack white patch; long, broad tail tipped whitish. Juvenile densely spotted below.
Voice Characteristic song is abrupt and loud, but far less melodious and varied than that of Northern Mockingbird, and not incorporating mimicry. Call is a *tyerrp*, longer and slightly higher pitched than Northern Mockingbird.
Similar species Northern Mockingbird is slightly smaller with conspicuous white wing-patch and unstreaked underparts; tail shorter with white borders, frequently held erect.
Habitat and behaviour Sandy coast vegetation complex, coastal thickets. Breeds April–July; builds nest among tall bushes; usually lays three creamy-white, spotted eggs (not described on Cuba). Feeds on insects, small fruits.

Juvenile.

Adult has long, wide tail and dark streaks on sides of whitish underparts.

Range Bahamas, Cuba and Jamaica.
Status *M. g. gundlachii* is rare and locally endangered due to habitat destruction. It is a breeding resident on some northern cays such as Paredón Grande, Cayo Guillermo and Cayo Cruz.

Cedar Waxwing

Bombycilla cedrorum 18cm

Adult.

Local name Picotero del Cedro.
Taxonomy Monotypic.
Description Head with conspicuous crest, cinnamon-brown crown, back and breast; black mask from base of bill to nape outlined with white; rump and uppertail pale grey. Long wings have grey-brown coverts with dark flight feathers and red shaft extensions at tip of secondaries. Pale yellowish abdomen, white undertail-coverts, short tail with yellow terminal band, bill blackish, legs dark brown. Juvenile has streaked buffy underparts.
Voice Hissing *szeeee* by flock on take-off.
Similar species None.
Habitat and behaviour Open woodland, pine forest and urban areas. Irruptive migrant, flocks coinciding with abundance of fruiting trees; also takes insects.

Adult has crest, black mask, red shaft extensions and a yellow-tipped tail.

Range Breeds in North America; winters in North America south through Middle America to Panama, and irregularly in the Bahamas, Cuba and Cayman Islands.
Status An erratic (less than annual) visitor, but sometimes reasonably common on passage, and less frequent winter visitor, on Cuba, Isle of Pines and the northern cays, October–May.

Blue-winged Warbler
Vermivora cyanoptera   12cm

Local name Bijirita de Alas Azules.
Taxonomy Monotypic.
Description Male has bright-yellow head (forehead to mid-crown), throat and underparts, with a distinct short black eye-line; nape and back olive-greenish; blue-grey wings with bold white wingbars; white vent and undertail-coverts. Female is duller, crown olive-green, eye-line blackish, and duller yellow underparts. Both sexes show white spots on tail. First-year male is similar to adult female. First-year female is duller with greenish crown, not contrasting with nape and upperparts.
Voice Sharp, loud *jeet*. Primary song a buzzy *beeee-buzzzz*.
Similar species Prothonotary Warbler lacks wingbars and black eye-line.
Habitat and behaviour Swamp woodland, riparian woodland, urban areas. Feeds on insects. Usually solitary. High site fidelity. Flicks tail.

Adult male.

Male. Note yellow crown and black eye-line, yellowish underparts, and bluish wings with two whitish wingbars.

Range Breeds in central-eastern United States; winters in Middle America, rarely to Panama, and migrates mainly along the Caribbean Gulf slope of Mexico. In the West Indies, an uncommon passage migrant and rare winter visitor in the Bahamas, Cuba and Hispaniola, and Cayman Islands.
Status Uncommon winter visitor and passage migrant on Cuba, Isle of Pines and some northern cays, August–April.

Golden-winged Warbler
Vermivora chrysoptera 13cm

Adult female.

Local name Bijirita Alidorada.
Taxonomy Monotypic.
Description The only warbler with yellow crown and large yellow wing-patch; blue-grey upperparts with white underparts; white undertail-coverts may have yellowish wash; and white spotted tail. Male has broad white supercilium, black throat and ear-coverts patch. Female has paler black on face and throat, crown greenish-yellow. Both sexes have extensive white on tail. First-year similar to adult.
Voice A loud sharp *jeet*.
Similar species Blue-winged Warbler is yellow below, with a black eye-line and white wingbars. Interbreeds with Golden-winged Warbler, resulting in hybrids with mixed plumage patterns.
Habitat and behaviour Open forests and semi-deciduous and evergreen woodland, swamp woodland.

Adult male has black throat and mask, and yellow crown and wing patch.

Range Breeds in North America; winters from Yucatán Peninsula to northern South America and rarely on passage in the West Indies.

Status Near Threatened. Very rare passage migrant on Cuba, September–October and February–May.

Tennessee Warbler

Leiothlypis peregrina 12cm

Local name Bijirita Peregrina.
Taxonomy Monotypic.
Description Non-breeding adult and breeding female have bright olive-green upperparts, yellowish-white supercilium, short blackish eye-line, fine straight bill, yellowish wash on breast and abdomen, darker flanks, white undertail-coverts, short tail with no tail-spots. Adults have faint yellowish wingbars, barely discernable when breeding. First-winter plumage similar to non-breeding adult with stronger yellowish wash on underparts; most have white undertail-coverts. Breeding male has blue-grey head, white supercilium and whitish underparts.
Voice Call *tssst*.
Similar species Philadelphia Vireo has heavier bill, bluish-grey (not blackish) eye-line; much more deliberate movements and is much less active while foraging.
Habitat and behaviour Semi-deciduous and evergreen woodland, second growth, riparian woodland and areas of scattered trees. The most active warbler. Feeds on insects, fruits and nectar.

Adult. Note white undertail-coverts.

Adult male has grey head, distinct blackish eye-line, and white supercilium and underparts.

Range Breeds in northern North America; winters from southern Mexico to northern South America, mainly on passage in the West Indies, and uncommon in winter, in the Bahamas, Greater Antilles east to Hispaniola, and Cayman Islands.

Status Common passage migrant and uncommon winter visitor on Cuba, Isle of Pines and many northern cays, and Cayo Grande off south coast, September–November and January–May.

Nashville Warbler

Leiothlypis ruficapilla 12cm

Adult female is duller than male.

Local name Bijirita de Nashville.
Taxonomy Polytypic (2).
Description Bluish-grey head and neck with concealed chestnut crown patch, olive-green back and wings, pronounced white eye-ring, no wingbars, underparts yellow except for small white area across lower abdomen. Unmarked tail. Female has smaller crown patch and duller yellow underparts. Both sexes in winter are duller.
Voice Call *jeet*.
Similar species None.
Habitat and behaviour Semi-deciduous and evergreen woodland.

Adult male. Note grey head, bold white eye-ring, and yellow underparts with whitish vent.

Range Breeds in North America; winters along the Gulf coast and in Middle America south to Guatemala; in the Bahamas and Cuba; vagrant elsewhere.

Status Rare winter visitor on Cuba, Cayo Coco and Cayo Paredón Grande, October–March.

Northern Parula

Setophaga americana 10.5–12.0cm

Local name Bijirita Chica.
Taxonomy Monotypic.
Description Adult has blue-grey head, back, wings and tail; greenish mantle patch, broad white wingbars, white broken eye-ring, yellow throat and breast; rest of underparts white; short tail with small white spots; yellow feet. Breeding male has plumage brighter, blackish lores and two bands across golden-yellow breast forming a necklace (upper band blue-grey and lower chestnut); female has grey lores and rufous breast-band is paler, a faint wash or absent. First-winter male resembles adult female.
Voice Call a sharp *chip*; song a short rising trill ending suddenly, heard occasionally in spring and early autumn.
Similar species None.
Habitat and behaviour Semi-deciduous and evergreen woodland, tropical karstic woodland, second growth, coastal thickets, pine woodland, riparian woodland and scattered trees. Gleans insects frequently under leaves in an upside-down position.

Female.

Non-breeding male has reduced breast-band.

Breeding male has two bands on breast.

Range Breeds in eastern North America south to northern Florida; winters in Florida, Middle America to Panama and mainly in the West Indies, in the Bahamas, Greater Antilles, Cayman Islands and northern Lesser Antilles.

Status Very common winter visitor and passage migrant throughout the Cuban archipelago, July–May.

Yellow Warbler

Setophaga petechia 11.5–13.5cm

Male has rufous streaks. Note the unique yellow tail-spots.

Juvenile is yellowish with faint streaks on underparts, a grey head and white throat.

Local name Canario de Manglar.
Taxonomy Polytypic (43).
Description Adults are yellow overall. Resident male has yellow-orange head with inconspicuous tawny-reddish cap; back, wings and tail olivaceous; wingbars and inner webs of tail feathers yellow; underparts yellow-orange streaked with rufous and grey-pink legs. Female (resident) is similar, but with olivaceous head and plain yellow underparts with faint tan streaks. Immature is entirely grey, with yellow in primary edges and tail; may have white throat. Wintering birds from North America appear virtually all yellow and lack the contrasting olivaceous and orange tones of residents.
Voice Song a loud, emphatic clear *tseeet tseeet tseeet ze-ze-ze tsweet*; call *chip*; also *zee-zee-zee*.
Similar species Female Hooded Warbler has white tail-spots, unmarked underparts and wings. Female Wilson's Warbler has unmarked tail and wings, and plain yellow underparts.
Habitat and behaviour Residents in coastal vegetation, especially mangroves; passage individuals prefer woodlands and urban areas. Gleans for insects and takes fruits. Pair bonds maintained and birds hold territories throughout most of year. Breeds March–July; lays 3–4 eggs, greenish-white spotted with brown, in deep cup nest of grasses and spiders' webs, lined with palm fibres; female broods and both parents feed young, which fledge in around 14 days.

Female has duller yellow underparts with faint streaking.

Adult male has orange streaks on breast and sides.

Range Breeds in North, Middle and northern South America. Fourteen endemic subspecies are resident in the West Indies. North American populations are migratory (*aestiva* group), observed in late winter and spring passage.
Status *S. p. gundlachi* is a common breeding resident throughout the Cuban archipelago. *S. p. aestiva* is fairly common passage migrant, August–November and February–May.

Chestnut-sided Warbler
Setophaga pensylvanica   11.5–13.5cm

Local name Bijirita de Costados Castaños.
Taxonomy Monotypic.
Description Non-breeding adult and first-winter have bright yellow-green upperparts, often with faint streaks, yellowish wingbars, white tail-spots, prominent white eye-ring, grey face and neck, and distinct chestnut sides (absent on immature female, reduced in adult female) on whitish-grey underparts. Breeding male has bright yellow crown, heavy streaks on back, black eye-stripe, nape and long malar stripe; white cheeks, sides of neck, wide chestnut band on sides and flanks; female duller with grey on face and chestnut on sides alone. White undertail-coverts in all plumages.
Voice Usually silent; call a loud *tsik*.
Similar species Female Blackpoll Warbler is more or less streaked above and below, appears more olive and yellow than green and white. Bay-breasted Warbler has broader white wingbars and buffy undertail-coverts.
Habitat and behaviour Semi-deciduous and evergreen woodland, from mid-level to canopy. Gleans insects on foliage.

Breeding male. In breeding, both sexes have yellow crown, black markings on head, chestnut on sides and white underparts – all of which are diagnostic of this species.

Non-breeding adult male. Non-breeding female has less chestnut on sides compared to males, or it is entirely absent.

Range Breeds in eastern North America; winters in southern Middle America to north-west South America; on passage through the West Indies.

Status Uncommon passage migrant on Cuba and some cays, September–November and February–May. One winter record on Cayo Coco.

Magnolia Warbler

Setophaga magnolia 11.5–12.5cm

First-year male has greyish breast-band; streaking on upperparts and sides less distinct. Note the pointed retrices.

Local name Bijirita Magnolia.
Taxonomy Monotypic.
Description In all plumages, tail pattern is diagnostic, formed by large white spots in centre of all but central feathers and broad black terminal band. Non-breeding adult has grey head, white eye-ring, two narrow white wingbars, bright yellow throat (unstreaked), breast and abdomen; black streaks on sides and flanks heaviest on male, reduced and blurred on female; yellow rump, white undertail-coverts. First-winter plumage has white eye-ring, streaking on greenish upperparts and pale grey band across upper breast. Both breeding adults have broad white supercilium; male has black face mask and back, white wing-coverts patch, and black breast-band joins broad black streaks on sides and flanks; female has white wingbars, dark grey mask, greyish-olive back streaked black, and streaking on yellow breast less bold. Frequently fans tail.
Voice Call a slurred repeated *tzek*.
Similar species None.
Habitat and behaviour Semi-deciduous and evergreen woodland, swamp woodland, urban areas. Gleans arthropods on outer branches of trees at low to mid-levels.

Breeding male. Note diagnostic broad black terminal band on tail.

Non-breeding adult.

Range Breeds in North America; winters mainly in Middle America casually to Panama, and to the West Indies in the Bahamas, Cuba, Hispaniola, Jamaica and Cayman Islands.

Status Common winter visitor and passage migrant to Cuba, Isle of Pines, many northern cays, and Cayo Cantiles and Cayo Largo off south coast, September–May.

Cape May Warbler
Setophaga tigrina 12.5–14.0cm

Local name Bijirita Atigrada.
Taxonomy Monotypic.
Description Small, with heavily streaked underparts, short tail and small bill. Non-breeding male has greyish-olive back streaked blackish, grey ear-patch often with rufous wash, yellow supercilium and patch on sides of neck, yellow underparts and breast heavily streaked, white undertail-coverts, large white wing-patch, greenish rump, white tail-spots. Female duller, with white wingbars, greyish ear-coverts, pale yellowish on sides of neck, thin streaking on breast. First-winter male has yellow supercilium and cheek-patch, lacks chestnut; female is drab with greyish-brown upperparts, yellow mostly absent with blurred olive streaking on buffy-grey underparts. Breeding male has bright chestnut ear-coverts, blackish crown, back heavily streaked black, and bright yellow rump; female has faint wingbars, greyish ear-coverts and pale yellow on face, yellow underparts (rump greenish-yellow) with narrower streaking.
Voice Call high-pitched *seet*.
Similar species Yellow-rumped Warbler has yellow patches on sides and brighter yellow rump in all plumages. Palm Warbler has yellow undertail-coverts and wags tail.
Habitat and behaviour Semi-deciduous and evergreen woodlands, swamp woodland, coastal thickets, urban areas. Gleans in the canopy of forest and bushes.

Breeding male has bright chestnut ear-coverts, yellow neck and dark streaking on back and underparts.

Non-breeding adult male. Note faint cheek-patch and reduced wingbar.

First-winter female is dull greyish-olive with blurred streaking on underparts, and faint yellow markings on face, neck and breast.

Non-breeding adult female.

Range Breeds in North America; winters casually in Florida and Middle America, and mainly in the West Indies (Greater Antilles) but uncommon in the Lesser Antilles.

Status Common winter visitor and passage migrant throughout the Cuban archipelago, September–May.

Black-throated Blue Warbler
Setophaga caerulescens 12–14cm

Local name Bijirita Azul de Garganta Negra.
Taxonomy Polytypic (2).
Description All plumages show a diagnostic small white rectangle on centre edge of closed

Adult female has long whitish supercilium and lower eye crescent; the only warbler with white rectangle on wing.

wing, reduced or absent in immature female; white undertail-coverts and small white tail-spots. Male has dark blue upperparts, black cheeks, throat and sides, white underparts. Female and first-winter are dark olive-brown above, pale brown below with narrow whitish eyebrow, white lower eye crescent and brownish-grey cheek patch; wing-spot smaller than in male. Immature male has green-washed upperparts and white-tipped throat feathers. Juvenile female has wing-patch indistinct or occasionally lacking.
Voice Call emphatic low *tik*.
Similar species No other warbler has a white patch on closed wing.
Habitat and behaviour Semi-deciduous and evergreen woodland, swamp woodland, tropical karstic woodland, cloud forest and urban areas. Feeds on insects and fruits from ground level to high in canopy, but most often seen on lower branches. Most abundant in high mountains.

Male has blue upperparts and a black face, throat and sides.

Range Breeds in eastern North America; the majority winter in the West Indies, in the Bahamas, Greater Antilles and Cayman Islands.
Status *S. c. caerulescens* is very common winter visitor and passage migrant throughout the Cuban archipelago, August–May. *S. c. cairnsi* is fairly common on Cuba and northern cays, October–April.

Yellow-rumped Warbler
Setophaga coronata 14cm

Local name Bijirita de Rabadilla Amarilla.
Taxonomy Polytypic (5).
Description Yellow rump and sides of breast, and white throat in all plumages. Non-breeding male has greyish-brown streaked upperparts, small yellow patch on crown and sides of breast, faint wingbars, whitish supercilium and prominent broken eye-ring. White below with blackish streaks on breast, variable greyish-brown streaks on sides, whitish undertail-coverts and extensive white on outer tail; female duller with yellow on sides reduced. First-winter similar but even duller, head and upperparts brownish, streaking on underparts blurred, yellow may show on rump alone. Breeding plumage seldom observed: male has black forecrown, lores and ear-coverts, blue-grey back with black streaks, yellow on cap and patch on sides, white underparts with black breast-band and bold streaking on sides; female similar but back brown, face greyish and reduced streaking on underparts.
Voice Call sharp *chep*.
Similar species Immature Cape May Warbler has diffuse yellowish-green rump, much less sharply defined, and shorter tail. Palm Warbler has prominent yellow undertail-coverts and wags tail frequently. Magnolia Warbler is mostly yellow below and tail has bold black terminal band.
Habitat and behaviour Open country, pine forest, urban areas. Feeds on insects.

First-winter male and female are indistinguishable in the field, with little (or no) yellow on sides, compared with non-breeding adult female.

Non-breeding adult female appears duller.

Non-breeding male has small yellow patches on crown, breast and sides.

Range Breeds in North America; winters in south-eastern United States, Middle America to Panama, and the West Indies where it is common in the Bahamas, Cuba, Jamaica and Cayman Islands, and irregular elsewhere.

Status *D. c. coronata* is a rather common winter visitor and passage migrant on Cuba, Isle of Pines, larger northern cays, and Cayo Cantiles and Cayo Largo off south coast, September–May.

Black-throated Green Warbler

Setophaga virens 12.5cm

First-winter has pale chin and throat and bold black streaks on sides and flanks.

Local name Bijirita de Garganta Negra.
Taxonomy Monotypic.
Description Male has bright olive-green upperparts, distinct yellow cheeks; chin, throat and upper breast black, sides streaked with black. White wingbars, white undertail-coverts with yellow wash across vent, white tail-spots. Female duller, with yellow chin and throat. First-winter male resembles adult female; first-winter female is duller still, without dark markings below apart from faint streaks on sides.
Voice Short, crisp *tsip*.
Similar species Townsend's Warbler (vagrant) has black crown and ear-covert patch. Female and immature Blackburnian Warbler lack black markings on throat and breast, have central dark area on cheeks and streaking on back darker.
Habitat and behaviour Semi-deciduous and evergreen woodland, swamp woodland, urban areas. Feeds on insects.

Female in spring, transitioning into breeding plumage, has pale chin, black on throat mixed with white, and black streaking on sides and flanks.

Range Breeds in North America; *S. v. virens* winters mainly from southern Florida through Middle America to Panama (uncommonly to northern South America) and the West Indies, particularly the Bahamas and Greater Antilles.
Status Common winter visitor and passage migrant on Cuba, Isle of Pines and some larger cays, August–May.

Blackburnian Warbler
Setophaga fusca 13cm

Local name Bijirita Blackburniana.
Taxonomy Monotypic.
Description All plumages have bold white wingbars (forming broad white patch on breeding males) and dark triangular ear-patch, black in males and greyish is females. Non-breeding male has blackish crown with yellowish-orange forecrown patch, chin, throat and wide supercilium curving around ear-patch to sides of neck, becoming intensely orange when breeding; blackish upperparts with pale stripes; abdomen creamy-white, streaked with black on sides; white undertail-coverts and outer tail feathers. Breeding female is duller, with yellow replacing orange, white wingbars and olive-streaked back. First-winter male resembles adult female but with black eye-line, yellower throat and heavier streaking on back; first-winter female very pale, with face and throat buffy-white, and faint streaks on sides and flanks.
Voice Silent.

Similar species Yellow-throated Warbler has triangular black ear-patch bordered with white, and unmarked grey back.
Habitat and behaviour Semi-deciduous and evergreen woodland, urban areas. Forages at height.

First-winter has olive crown, wide yellowish supercilium, grey cheek and yellow throat.

Breeding male has orange on forehead and throat; black crown and ear-coverts patch.

Range Breeds in North America; winters from southern Middle America to north-western South America; autumn passage is mainly along the Atlantic coast and Florida.

Status Rare passage migrant on Cuba, Isle of Pines, many northern cays, and Cayo Largo and Cayo Caguama off south coast, August–December and February–May.

Yellow-throated Warbler

Setophaga dominica 13cm

Adult.

Local name Bijirita de Garganta Amarilla.
Taxonomy Polytypic (4).
Description Grey upperparts unstreaked, black on forehead and crown (variable) and triangular black ear-patch bordered with white, broad white supercilium and white wingbars. Some individuals may have yellow near lores. Throat and upper breast yellow, abdomen white, sides heavily streaked with black; bill long and tail with extensive white patches on outer rectrices. Sexes very similar, with female slightly duller with greyer crown. First-winter may be almost imperceptibly washed with brown on back. 'Sutton's' Warbler (a hybrid between Yellow-throated Warbler and Northern Parula) has been reported only once.
Voice A loud, dry *clip*.
Similar species Blackburnian Warbler has triangular black or dark ear-patch surrounded by orange or yellow. Olive-capped Warbler lacks white supercilium, has bright yellow throat and upper breast-patch bordered with black streaks, no streaking on sides.
Habitat and behaviour Semi-deciduous and evergreen woodland, swamp woodland, urban areas, coconut groves. Feeds on insects and spiders.

Adult. Note triangular black ear-patch, bordered white, broad white supercilium and yellow throat.

Range Breeds in south-eastern North America; winters in south-eastern United States, Middle America to Panama, the Bahamas, Greater Antilles and Cayman Islands.

Status *S. d. dominica* and *S. d. albilora* are common in winter and on passage on Cuba, Isle of Pines and some larger cays, July–April. *S. d. stoddardi* is a vagrant.

Olive-capped Warbler
Setophaga pityophila 13cm

Local name Bijirita del Pinar.
Taxonomy Monotypic.
Description Grey above, with olive-yellowish forehead to mid-crown and narrow white wingbars, yellow throat and breast bordered with small black streaks. Male brighter with bold blotchy black streaks; whitish abdomen sharply demarcated from yellow breast. Juvenile entirely brown with indistinct whitish wingbars; some individuals show yellowish cast on chin.
Voice Song, a series of 7–9 whistled watery notes, descending slightly and changing in quality. Call, a frequently heard series *tsip-tsip-tsip*.
Similar species Yellow-throated Warbler has dark crown and black streaks on flanks.
Habitat and behaviour Found only in pine forests from sea level to 850m. Breeds March–June. Nest cup-shaped, fairly high in pines and lays two whitish, spotted eggs. Feeds on insects.

Adult male. Note the bright yellow throat and breast, and well-marked blotchy black streaks on breast.

Adult. Note the olive-yellowish crown.

Range Grand Bahama, Abaco and Cuba.
Status Common but very local breeding resident on Cuba, in Pinar del Río (La Güira, Viñales Valley, San Ubaldo) and the eastern provinces of Holguín and Guantánamo.

Prairie Warbler

Setophaga discolor 13cm

First-winter male has blackish on face and flanks.

First-winter female has greyish face and ear-coverts.

Local name Bijirita Amarilla de Costados Rayados.
Taxonomy Polytypic (2).
Description Male is olivaceous above with chestnut streaks on back, yellowish wingbars, yellow face with black eye-line and crescent under eye; bold black streaks on sides. Female somewhat duller with less distinct black markings. Non-breeding male similar to breeding with less distinct streaks on side and head pattern, and less marked chestnut streaks on upperparts. First-winter male is similar to adult female but brighter, with more conspicuous black streaks on side and head pattern. First-winter female is greyish-olive above, with grey wash on head and much reduced facial contrast and streaks on sides, and yellowish underparts. Wags tail.
Voice A dry, husky *chip*; an ascending, thin buzzy song *zee, zee, zee, zee, zee, zee, zee, zee, zee*.
Similar species Palm Warbler is greyish-brown above with whitish supercilium and faintly streaked whitish underparts with contrasting yellow undertail-coverts. Pine Warbler has white abdomen and lacks black facial streaks. Winter Magnolia Warbler has greyish-olive back streaked with black, yellow rump, and black-tipped tail with white central band.
Habitat and behaviour Open forests, coastal thickets, urban areas. Feeds on insects, spiders, fruits mostly from small bushes to mid-height trees.

Adult female has duller plumage. Note greyish-olive supercilium and curve under eye and reduced streaking on sides.

Adult male has rufous streaks on back, and black streaks on sides and flanks.

Range Breeds in eastern North America; *S. d. discolor* winters mainly in south-eastern United States, the Caribbean coast of Middle America and the West Indies.

Status *S. d. discolor* is a common winter visitor and passage migrant throughout the Cuban archipelago, July–May. *S. d. paludicola* is rare winter visitor and passage migrant.

Palm Warbler

Setophaga palmarum 12.5–14.0cm

Local name Bijirita Común.
Taxonomy Polytypic (2).
Description Yellow undertail-coverts are diagnostic in all plumages. Non-breeding adult and first-winter have brownish crown and upperparts, white tail-spots, long pale supercilium, long dark eye-line, greenish-yellow rump, underparts greyish-buff with faint grey streaks. Breeding adult of western subspecies has chestnut crown, bright yellow supercilium, throat and upper breast with rufous streaks. Eastern subspecies has all-yellow underparts.
Voice A distinct frequent *chip*.
Similar species First-winter Cape May Warbler has white undertail-coverts and is more heavily streaked below. Yellow-rumped Warbler has white undertail-coverts and bright yellow rump.
Habitat and behaviour Second growth, open semi-deciduous and evergreen woodland, swamp woodland, tropical karstic woodland, pine forest, grassland, urban areas. Bobs tail continually. Mainly terrestrial, forages for insects.

Breeding adult has chestnut crown, yellow supercilium, dark eye-line and soft, grey streaks on underparts.

Non-breeding adult lacks rufous cap and streaking on underparts is less prominent. Note yellow undertail-coverts in all plumages.

Range Breeds in North America; *S. p. palmarum* winters mainly in Florida and the West Indies and sparsely along the United States Gulf coast and in Middle America.

Status Very common winter visitor and passage migrant throughout the Cuban archipelago, August–May. *S. p. hypochrysea* is a very rare winter visitor.

Bay-breasted Warbler

Setophaga castanea 12.5–15.0cm

Local name Bijirita Castaña.
Taxonomy Monotypic.
Description All plumages have two broad white wingbars, buffy undertail-coverts and extensive white tail-spots; dark legs and feet. Non-breeding male has greenish-olive crown and back with heavy blackish streaks (may be absent); obscure supercilium, grey eye-line, pale buffy-white below, with extensive chestnut wash on sides and flanks. Non-breeding female is similar to male but crown and upperparts faintly streaked (always lacking chestnut on head); sides warm buff, sometimes with chestnut feathers. First-winter male is duller than adult female. First-year female is very dull, olive-yellow above, with very faint streaks on underparts; whitish or yellowish-buff underparts and warm buff on sides lacking. Breeding male has heavily-streaked back, black forehead and face; cream patch on side of neck to hindcrown, dark chestnut crown; chin, throat, sides and rest of underparts whitish.
Voice A sharp *jeet*.
Similar species Non-breeding and immature Blackpoll Warbler slightly darker, upperparts are more olive-green with streaked yellowish underparts, less broad and contrasting wingbars; extensive pure white undertail-coverts are long, making tail appear short; dark or yellowish legs and feet with yellowish soles. Non-breeding Chestnut-sided Warbler has bright unstreaked green upperparts, conspicuous white eye-ring and undertail-coverts, yellowish wingbars.
Habitat and behaviour Semi-deciduous and evergreen woodand. Forages high in trees.

Breeding adult male.

Breeding adult male.

Non-breeding adult male has streaking on crown and back, white bar across wing and faint chestnut on flanks.

Breeding female has large buffy patch on neck, and chestnut on crown and upper breast side.

Range Breeds in North America; winters from Panama to Colombia and north-western Venezuela. Migrates primarily across the Gulf in spring and via the Atlantic route in autumn, with low numbers on passage in the West Indies, in the Bahamas, western Greater Antilles and Cayman Islands; vagrant elsewhere.
Status Rare but regular passage migrant on Cuba, Isle of Pines and many cays, mainly off the north coast, September–November and March–May.

Blackpoll Warbler

Setophaga striata 14cm

Local name Bijirita de Cabeza Negra.
Taxonomy Monotypic.
Description All plumages have heavily streaked back, very distinct white wingbars, white tail-spots and long white undertail-coverts. Non-breeding adult and first-winter plumages have greenish-olive upperparts, throat and breast yellowish-buff with faint streaks, lower abdomen white; soles of feet, and sometimes legs, yellowish, otherwise greyish. Breeding male has black cap to nape and black malar stripe, white cheeks, back streaked black, white underparts with black streaks from neck to flanks, and orange-yellow legs. Breeding female has short whitish supercilium, whitish face faintly mottled darker, crown and upperparts olive-grey with black streaks, dark eye-line, light streaking on whitish underparts, legs similar to male but usually darker on sides. First-year is duller, with indistinct streaks on underparts and blurred olive streaks on breast sides.
Voice Usually silent, call clear *chip*.
Similar species Black-and-white Warbler is black above, with narrow white stripes on head and back. Non-breeding Bay-breasted Warbler is yellowish-green above with bolder contrasting broad wingbars (darker area between bars), buff undertail-coverts and usually unstreaked underparts; short undertail-coverts make tail appear long; legs and feet grey.
Habitat and behaviour Coastal thicket, rocky and sandy vegetation complex, often on or near ground or low in bushes or trees.

Breeding male has black cap, white cheeks, black malar stripe, streaked back and sides, white wing-bars and yellow-orange legs.

Breeding female has streaked crown and back, whitish face, dark eye-line, streaking on sides and yellow legs.

Range Breeds in North America; winters in north-east South America to northern Bolivia and northern Brazil. Migrates almost exclusively through the West Indies, on Cuba in autumn over the central and eastern parts of the island, and in spring through the Bahamas, western Greater Antilles and Cayman Islands.
Status Common passage migrant throughout the Cuban archipelago, August–December and March–June. Two winter records on Cuba.

Cerulean Warbler
Setophaga cerulea 12cm

Adult female.

Local name Bijirita Azulosa.
Taxonomy Monotypic.
Description The only warbler in which both sexes show blue upperparts, paler in females. Male has bright cerulean-blue upperparts, streaked back, bold white wing-bars, white underparts with black narrow breast-band and streaking on sides, and a rather short tail with white spots. Female is turquoise-blue above with whitish-buff supercilium broadening behind eye, bold white wing-bars, and faint streaking on sides. First-year male has bluer upperparts than female, especially on rump; heavily streaked back. First-year female has greener upperparts, and yellowish supericulium and undperarts.
Voice A thin *chip*.
Similar species None.

Male has blue upperparts with white underparts, a black breast-band and bold black streaks.

Range Breeds in North America; winters in northwestern South America.
Status Vulnerable. Very rare passage migrant, August–November and April.

NEW WORLD WARBLERS 291

Black-and-white Warbler

Mniotilta varia   12.5–14.0cm

Local name Bijirita Trepadora.
Taxonomy Monotypic.
Description The only warbler boldly streaked in black and white. Non-breeding male has black-and-white striped crown, face and streaked upperparts, broad white supercilium, white wingbars, small tail-spots on outer rectrices, undertail-coverts white with black spots; variable amounts of white on throat and underparts, bold black streaks on sides and flanks and blackish ear-coverts. Female duller, grey ear-coverts, and buffy flanks with blurred narrower streaking. Breeding male is brighter with black throat; female more white than black, ear-coverts buffy-grey, white throat, buffy-white underparts with greyish streaking on sides and buffy flanks. First-winter plumage resembles female but throat often white and streaking grey; underparts buff in female.
Voice Usually silent, occasionally fast *chip-chip* call.
Similar species None.

Habitat and behaviour Semi-deciduous and evergreen woodland, tropical karstic woodland, swamp woodland, pine forest, rainforest, urban areas. Creeps around long branches and up and down the boles of trees foraging for insects.

Non-breeding male has black ear-coverts patch and throat mottled whitish.

Breeding female has white throat, black eye-line, grey cheek and blurred streaking below.

Range Breeds in North America; winters from Florida, Gulf coast, Middle America to northern South America, and the West Indies.
Status Very common winter visitor and passage migrant throughout the Cuban archipelago, July–May.

American Redstart

Setophaga ruticilla 11.0–13.5cm

First-year female has duller yellow patches on breast; wings have little or no yellow.

Local name Candelita.
Taxonomy Monotypic.
Description Male has black head, upperparts, throat and breast; brilliant orange patches on wings (base of primaries in flight), sides of breast and base of outer tail; white abdomen and undertail-coverts. Female has grey head, olive-grey upperparts, whitish supercilium, eye-ring and lores; orange patches replaced by lemon-yellow. First-year male resembles adult female with breast sides usually orange-yellow, and in first summer black appears on lores, head and breast, and yellow deepens to orange. First-year female is duller than adult, with grey head, pale yellow on sides and tail, often absent on wings.
Voice Call soft high *chit*. The song, commonly heard on Cuba, is a quick series of five or six notes, usually with a slurred conclusion.
Similar species None.
Habitat and behaviour Occurs from sea level to mid elevations (at least 1,300m in eastern third of island) in semi-deciduous woodland and second growth, riparian woodland and tropical karstic woodland, scattered trees in urban areas. Very active when foraging; fans tail, takes mainly insects by hover-gleaning and aerial fly-catching. Males hold winter territories.
Breeding Two reports on Cuba in 1989 and 1990.

Male is black with orange areas on side of breast, wings and tail.

First-year male has black blotching on throat and breast.

Range Breeds in North America, including northern Florida and the Gulf coast; winters in central Florida, and from southern Mexico to north-west South America and the northern West Indies.

Status Very common winter visitor and passage migrant throughout the Cuban archipelago, July–May.

Prothonotary Warbler
Protonotaria citrea   13.5cm

Local name Bijirita Protonotaria.
Taxonomy Monotypic.
Description Male has intensely orange-yellow head, throat and breast, olive back, and long black bill and eye; abdomen paler-yellow, with white vent and undertail-coverts. Female slightly duller, especially on head; crown and nape washed olive; white mid-abdomen to undertail-coverts. Both sexes have unmarked bluish-grey wings and short, bluish-grey tail with large white spots. Lower belly and undertail-coverts white. First-year male is similar to adult female with large spots on tail. First-year female duller than adult female with crown and nape heavily washed olive; tail has smaller tail-spots.
Voice Loud *chip*.
Similar species Blue-winged Warbler has white wingbars and black eye-line.
Habitat and behaviour Semi-deciduous and evergreen woodlands, swamp woodland, riparian woodland, urban areas, mangrove woodland; regularly forages at medium height near streams.

Female has duller head, and yellow throat and breast.

Male has golden-yellow head, throat and breast.

Range Breeds in eastern North America to the Gulf coast and central Florida; winters from the Yucatán Peninsula to northern South America; on passage in the West Indies.

Status Uncommon passage migrant on Cuba, Isle of Pines and some northern cays, August–November and February–April.

Worm-eating Warbler
Helmitheros vermivorum 14cm

Adult.

Local name Bijirita Gusanera.
Taxonomy Monotypic.
Description The only warbler with four distinct black stripes on head. Olive-brown back, cinnamon-buff crown with black lateral stripes and black eye-line to nape; cinnamon-buff throat and breast, undertail-coverts buffy-white; large pale bill, and pinkish legs; no wingbars or tail-spots. Sexes alike.
Voice Call loud *thip*.
Similar species Swainson's Warbler has dark flat brown crown and long creamy supercilium, normally found on or near the ground.
Habitat and behaviour Semi-deciduous and evergreen woodland, coastal thickets. Rummages noisily for insects and spiders among hanging dry leaves often high in trees.

Adult. Four blackish stripes on cinnamon-buff head is diagnostic of this species.

Range Breeds in eastern North America; winters on the Caribbean slope from central Mexico to Panama and, in lower numbers, in the West Indies.

Status Uncommon winter visitor and passage migrant throughout the Cuban archipelago, August–May.

Swainson's Warbler

Limnothlypis swainsonii   14cm

Local name Bijirita de Swainson.
Taxonomy Monotypic.
Description Brownish-olive back, wings and short tail, warm brown sloping forehead, crown and nape, long wide creamy supercilium, long dark eye-line, buff-white to yellowish-white underparts, washed olive-grey on sides; long pointed pale bill, wider at base, and pale pinkish legs; no wingbars or tail-spots. First-year has more pointed retrices.
Voice Short, metallic, *ziip* often difficult to locate, also *sreee* flight call, and liquid *chip* not unlike that of American Redstart.
Similar species Worm-eating Warbler has head stripes and usually feeds high in trees.
Habitat and behaviour Inhabits very humid semi-deciduous and evergreen woodland with abundant bromeliads and leaf litter, secondary woodland and swamp woodland, from sea level to mid-elevations. Very secretive. Shows high site fidelity.

Adult. Note long, wide-based bill, warm brown cap and wide creamy supercilium.

Range Breeds in the south-eastern United States to northern Florida; winters in the Yucatán Peninsula and Belize, and the West Indies in the Bahamas, Greater Antilles west of Hispaniola and Cayman Islands.

Status Fairly common winter visitor and passage migrant on Cuba, Isle of Pines and mainly northern cays, and Cayo Real off south coast, September–April.

Ovenbird

Seiurus aurocapilla   14.0–16.5cm

Local name Señorita de Monte.
Taxonomy Polytypic (3).
Description Olive-brown upperparts, orange crown bordered by black stripes, distinct broad white eye-ring, black malar stripe, white underparts with throat and breast boldly streaked with black; legs pink. Immature similar to adult, but with rusty-edged tertials (usually visible only in hand). No wingbars or tail-spots.
Voice Call *chuup*.
Similar species None.
Habitat and behaviour Semi-deciduous and evergreen woodland, tropical karstic woodland, wetlands, typically with abundant leaf litter. Ground dweller, thrush-like. Bobs head and carries tail erect with wings drooped while foraging for insects by flipping leaves. Solitary, holding winter territory.

Adult. Note orange crown bordered by lateral stripes, and pale legs. Tail is often cocked.

Range Breeds in North America; winters from southern United States to northern South America and the West Indies.
Status *S. a. aurocapilla* is common winter visitor and passage migrant throughout the Cuban archipelago, August–May. The subspecies *S. a. furvior* is rare in passage.

Northern Waterthrush

Parkesia noveboracensis 12.5–15.0cm

Local name Señorita de Manglar.
Taxonomy Monotypic.
Description Head and upperparts dark olive-brown, broad long supercilium yellowish or whitish (not bicoloured) and narrows towards nape, underparts whitish or yellowish with dark streaks on throat (fine), breast (heavy) and sides; legs grey-pink. No wingbars or tail-spots.
Voice Loud *chink* lower pitched than Louisiana Waterthrush.

Similar species Louisiana Waterthrush is larger, white throat is usually unspotted, supercilium is buff in front of eye, white and broad behind, buffy flanks; less densely streaked underparts.
Habitat and behaviour Wetlands, forest edge (on migration), swamp woodland. Holds winter territory; constantly pumps rump and tail mostly downwards while walking on open ground or on low branches searching for arthropods.

Adult. Note supercilium that tapers behind eye towards nape, and heavily streaked underparts.

Range Breeds in northern North America; winters through Middle and South America to Peru, Brazil and the West Indies.

Status Common winter visitor and passage migrant throughout the Cuban archipelago, July–May.

Louisiana Waterthrush
Parkesia motacilla 15cm

Local name Señorita de Río.
Taxonomy Monotypic.
Description Head and upperparts dark olive-brown, broad long supercilium buff in front of eye, white and widening behind to nape, dark eye-line; throat usually unspotted, white undertail-coverts (sometimes with buffy wash), underparts white with brown streaks on breast, sides and buffy flanks, long pointed bill pale at base, bright pink legs; no wingbars or tail-spots.
Voice Loud *chink* higher pitched than Northern Waterthrush.
Similar species Northern Waterthrush is smaller, supercilium is yellowish or whitish tapering towards nape, throat spotted, underparts yellowish or white with dark dense streaks on breast.
Habitat and behaviour Mainly terrestrial, often in more open habitats than Northern Waterthrush; sometimes forages at edge of lagoons and ponds; usually near flowing water. Moves tail slowly up and down and from side to side continually.

Adult has long supercilium that becomes broader behind eye.

Range Breeds in eastern North America excluding Florida; winters through Middle and northern South America, West Indies, and especially the Greater Antilles.
Status Common winter visitor and passage migrant on Cuba, Isle of Pines, Cayo Las Brujas and Cayo Coco, July–April.

Kentucky Warbler
Geothlypis formosa 12.5–14.5cm

Male has black area from lores to cheek and side of neck.

Local name Bijirita de Kentucky.
Taxonomy Monotypic.
Description All plumages have blackish crown, yellow 'spectacles', bill pale at base, legs pinkish, greenish-olive upperparts, yellow underparts and undertail-coverts. No wingbars or tail-spots. Male has more distinct black forehead and crown, black triangular patch from lores to below eye and sides of neck. Female and first-winter male have reduced black markings on crown and sides of neck. First-winter female has olive crown and flanks, dark grey ear-coverts and sides of neck.
Voice Usually silent, call *tik*.
Similar species Immature male Common Yellowthroat lacks yellow spectacle, and never appears entirely bright yellow below.
Habitat and behaviour Semi-deciduous and evergreen woodland.

Female has reduced black markings on crown and neck. Both sexes have yellow 'spectacles' and underparts.

Range Breeds in south-eastern United States excluding Florida and the Gulf coast; winters mainly from Mexico to Panama and northern South America, rarely in winter and on passage in the West Indies.

Status Rare passage migrant and very rare winter visitor on Cuba, Isle of Pines and some northern cays, August–April.

Common Yellowthroat

Geothlypis trichas 11.5–14.0cm

Local name Caretica.
Taxonomy Polytypic (13).
Description All plumages have brownish-olive upperparts, no wingbars or tail-spots, yellow throat and yellow-olive undertail-coverts, sides and flanks washed brownish. Adult male has wide black mask bordered above by ashy-grey; bright yellow throat and breast. Female duller, lacks mask, brownish ear-coverts, less bright yellow throat and breast. First-winter male has some blackish on face; female drab, with pale buff eye-ring, buffy or pale yellowish throat and breast; undertail-coverts have yellowish wash, and brown wash on flanks.
Voice Call grating *chack*.
Similar species Kentucky Warbler has yellow underparts in all plumages, with yellow 'spectacles'.
Habitat and behaviour Bushy tangles in semi-deciduous and evergreen woodland, tropical karstic woodland, riparian woodland, wetland and pine forest, coastal thickets.

First-year male transitioning to adult plumage.

Breeding male.

First-winter female has yellowish wash on breast and pale eye-ring.

Adult female has faint supercilium and pale eye-ring.

Range Breeds throughout North America and northern Mexico; *G. t. trichas* winters from extreme southern United States to Middle America (casually to northern South America) and the West Indies.

Status Very common winter visitor and passage migrant throughout the Cuban archipelago, September–May.

Yellow-headed Warbler

Teretistris fernandinae   13cm

Local name Chillina.
Taxonomy Monotypic.
Description Adult has a conspicuous yellow head with distinct yellow eye-ring, plain grey back, wings and tail (both unmarked); greyish-white underparts. Bill thick and slightly decurved.
Voice Very noisy, rasping *shhh-shhh-shhh-shhh shhh-shhh*.
Similar species Oriente Warbler has entirely grey upperparts, extensively yellow underparts.
Habitat and behaviour Semi-deciduous woodland, swamp woodland, tropical karstic woodland, pine forest, coastal thickets and sandy coast vegetation complex, from sea level to high elevations. Foraging habits are highly variable. Breeds March–July; builds cup-shaped nest of grasses, rootlets and other plant fibres. Lays 2–3 pale green eggs, spotted with lilac and reddish-brown at large end. Feeds in conspecific flocks among branches and foliage, on bark and on ground. Feeds on insects, small lizards, spiders.

Adult.

Adult. Note yellow head with distinct yellow eye-ring and slightly downcurved bill.

Range Cuba. Range overlaps with Oriente Warbler (hybrids reported).
Status Endemic species, common but restricted to western and central Cuba, Isle of Pines and Cayo Cantiles.

NEW WORLD WARBLERS 301

Adult.

Oriente Warbler

Teretistris fornsi 13cm

Adult.

Local name Pechero.
Taxonomy Monotypic
Description Upperparts plain grey, contrasting with bright yellow face, very distinct eye-ring and breast. Grey wings and tail unmarked. Bill thick and slightly decurved. In eastern populations, yellow expands to mid-abdomen; lower abdomen and undertail-coverts greyish.
Voice Noisy and rasping *shhh-shhh-shhh-shhh-shhh-shhh* not quite as harsh as that of Yellow-headed Warbler.
Similar species Female Prothonotary Warbler is olivaceous above, with white spots on tail. Yellow-headed Warbler has entirely yellow-olive head, and greyish chest and abdomen.
Habitat and behaviour Semi-deciduous and evergreen woodland, pine forest, coastal thickets and sandy coast vegetation complex, from sea level to highest elevations of Sierra Maestra and Nipe–Sagua Baracoa mountain range. Like Yellow-headed Warbler, has highly variable foraging habits. Forms small flocks outside breeding season and feeds in mixed flocks on insects, spiders and small lizards. Breeds March–July. Builds cup-shaped nest of grasses, rootlets and other plant fibres. Lays 2–3 very pale green eggs, spotted with lilac and reddish-brown at large end.

Adult. Note grey upperparts, yellow eye-ring, face, throat and breast, and long slightly downcurved bill.

Range Cuba.
Status Endemic species, common but restricted to central and eastern Cuba and the larger northern cays.

Adult.

Hooded Warbler

Setophaga citrina 12.5–14.5cm

Adult female has yellow face, edged blackish, yellow throat and dark lores.

Local name Monjita.
Taxonomy Monotypic.
Description Male has black hood (crown, nape and throat) enclosing golden-yellow forehead and face, greenish upperparts. Bright yellow underparts, black iris and pink legs, extensive white tail-spots on outer rectrices. Female usually has reduced hood with blackish cowl framing face, yellow throat and dark lores. Unmarked wings. First-year male similar to adult male with olive-yellow edges to hood; first-year female has olive-green crown, nape and upperparts; face and underparts yellowish.
Voice Usually silent, call emphatic ringing *chip*.
Similar species Female Wilson's Warbler has olivaceous sides of head, blending inconspicuously into olive back; lacks tail-spots.
Habitat and behaviour Semi-deciduous, evergreen woodland, swamp woodland. Forages from ground, where it hops about rapidly, to middle levels; often flicks and fans tail, flashing the white spots. Feeds on insects on the wing.

Adult male has black hood and golden forehead and cheek.

Range Breeds in North America to Gulf coast and northern Florida; winters mainly in Middle America (majority in south-eastern Mexico) to Panama and the West Indies.

Status Uncommon passage migrant and rare winter visitor, on Cuba and many cays, August–May.

Wilson's Warbler

Cardellina pusilla   12cm

Local name Bijirita de Wilson.
Taxonomy Polytypic (3).
Description Male has black cap, yellow face with distinct dark eye and short bill; greenish-olive upperparts with unmarked wings; yellow underparts, pale legs and long dark unmarked tail, often flitting or cocked. Female has olive crown with some black on forecrown. First-year male has olive feathers on black cap. First-year female lacks black markings on head.
Voice A husky *chuck*.
Similar species Female Hooded Warbler has large yellow patch on sides of face and flashes white spots on tail. Female Yellow Warbler is larger, with yellow edges to primaries and inner webs of tail feathers (tail-spots).
Habitat and behaviour Swamp woodland, coastal thickets and urban areas. Feeds on insects high in canopy. Frequently flicks tail upwards.

Adult male has black cap.

Range Breeds in North America; winters from Gulf coast through Middle America to Panama. Casual on passage and in winter in the Bahamas and Greater Antilles where rare on Jamaica, Cuba, and Cayman Islands.

Status *C. p. pusilla* is very rare passage migrant on Cuba, September–November and February–April. Multiple winter records, and some individuals may winter. One record of *C. p. pileolata*.

Canada Warbler

Cardellina canadensis   12.5–15.5cm

Local name Bijirita del Canadá.
Taxonomy Monotypic.
Description The only warbler with streaks on mid-breast. Dark blue-grey upperparts, bold white eye-ring, yellow above lores and underparts, no wingbars, white undertail-coverts, pale pinkish legs and long unmarked tail. Male has black forehead, forecrown and sides of throat, bold black streaked necklace across yellow breast. Female similar but duller, face and necklace grey and less distinct. First-winter similar with olive-grey upperparts and necklace faint or absent.
Voice Usually silent.
Similar species No other warbler has streaks restricted to the breast.
Habitat and behaviour Open woodland and scattered trees near flowing water. Feeds on insects.

Adult male has prominent black 'necklace' and bluish-grey upperparts.

Range Breeds in North America; winters in South America from Colombia to Peru and Brazil. On passage mainly through Middle America, rarely in the Bahamas, Cuba (autumn) and Cayman Islands; casual or vagrant elsewhere in the Greater Antilles.
Status Very rare passage migrant and winter visitor, September–October.

Yellow-breasted Chat

Icteria virens   19cm

Adult male has black bill and lores. Note white 'spectacles' and bright yellow throat and breast.

Local name Bijirita Grande.
Taxonomy Polytypic (2).
Description Large with olive upperparts; white eye-crescent, supraloral and submoustachial stripe, thick black bill with curved culmen, black lores, bright yellow-orange throat and breast, white abdomen and undertail-coverts; unmarked wings and long rounded tail. Female has grey lores and greyish-black bill with flesh-coloured mandible base.
Voice A single *kook*.
Similar species None.
Habitat and behaviour Coastal thickets.

Range Breeds in North America to central Mexico; winters mainly in Middle America to Panama. Rare on passage in northern Bahamas and Greater Antilles.

Status *I. v. virens* is rare passage migrant on Cuba and cays, October–November and February–May.

Bananaquit

Coereba flaveola   10.0–12.5cm

Local name Reinita.
Taxonomy Polytypic (41).
Description Black upperparts with long broad white supercilium, black eye-line to nape and yellow rump; yellow mid-breast to upper belly, throat and rest of underparts greyish-white; white spot on wing at base of primaries and yellow on bend of wing; thin black decurved bill with red spot on base. Juvenile has greyish-brown upperparts and buffy-grey underparts with patchy yellow; greyish-yellow supercilium.
Voice Call *tsiip*; grating varied song including *churr-rr-rr-rr* and level high *te-zi-te-zi-te-zi*.
Similar species None.
Habitat and behaviour Coastal thickets. Breeding has not been confirmed. Forages from canopy to close to ground, primarily on nectar (flower piercer), moving rapidly between flowers; also takes fruits and small insects.

Adult. Red spot on gape, white square on wing and decurved bill are all diagnostic of this species.

Juvenile has dull greyish plumage with faint buffy supercilium.

Range Mexico to northern Argentina and the West Indies.
Status Rare visitor, mainly on the northern cays.

Summer Tanager

Piranga rubra 18.0–19.5cm

Local name Cardenal.
Taxonomy Polytypic (2).
Description Adult male entirely red with underparts brighter; plumage retained in all seasons. Crown slightly crested, large stout bill, darker red wings and tail, legs grey. Female has rich yellow head and upperparts, whitish eye-ring, yellowish-orange underparts, some with reddish wash. First-winter male similar to female, becoming increasingly patchy red on head and breast in first spring.
Voice Call loud, rapid *pit-i-tuk*.
Similar species Scarlet Tanager female is greenish-yellow, non-breeding male similar but brighter green, wings and tail blackish not dark olive, and shorter bill.
Habitat and behaviour Semi-deciduous and evergreen woodland, swamp forests, coastal thickets.

Adult male is entirely red.

Adult male.

Adult female.

First-year male.

Range Breeds in United States and northern Mexico; winters from Middle America to South America and the West Indies.
Status *P. r. rubra* is uncommon winter visitor and slightly more common passage migrant on Cuba, Isle of Pines and northern cays, September–May.

Scarlet Tanager
Piranga olivacea 18cm

Local name Cardenal de Alas Negras.
Taxonomy Monotypic.

Breeding male is entirely red with black wings and tail.

Description Breeding male (in late spring) is bright scarlet with black wings and tail, small stout bill, grey legs. Female in all plumages has olive-green to greenish upperparts, yellowish underparts, dark olive wings and tail edged brownish; non-breeding male similar to female with olive-yellow underparts, blackish wings and tail with no olive edging on coverts. First-year male is greenish-yellow, becoming increasingly mottled with red.
Voice Silent.
Similar species Summer Tanager male is entirely red; female and immature male are more orange-yellow than greenish, wings and tail dusky, bill longer and heavier, axillaries yellow.
Habitat and behaviour Very stolid; semi-deciduous and evergreen woodland, swamp woodland.

Non-breeding male has black wings and tail. Female has dark olive wings and tail.

Range Breeds in North America; winters in South America and occurs on passage in Middle America and the West Indies.	**Status** Scarce, but regular, passage migrant throughout the Cuban archipelago, September–November and February–May.

Western Spindalis
Spindalis zena 15cm

Local name Cabrero.
Taxonomy Polytypic (5).
Description Male has greenish-yellow mantle, brownish-orange nuchal collar, rump and uppertail-coverts. Head is black with very bold white stripes; short, thick dark bill; white chin and throat, upper breast rich chestnut-brown (extent and density varies), yellow on lower breast; abdomen greyish. Black wings with broad white patch at base of primaries and extensive white edgings; tail black and outer rectrices edged white, legs grey. Female is dull, upperparts greyish with olive-green wash, indistinct whitish stripes on head, wings edged whitish or greenish with whitish square on closed wing, underparts greyish-white; females often have orange flush on shoulders, rump and upper breast. Juvenile male similar to adult female with yellow-orange wash and blackish streaks on back. Female duller, with square on wing usually absent.

Voice Song, a variable, jumbled series of high *tseeps* notes and buzzes, e.g. *weeze-weeze-weeze-tee-tee-bizz-weeze-weeze*. Most characteristic call is a thin, high-pitched *tsiiiiiii*; also a short, high *tseep*.

Similar species None.

Habitat and behaviour Open forests, semi-deciduous and evergreen woodland, swamp woodland, coastal thickets, pine forest, from sea level to highest altitudes. Breeds February–July. Builds cup-shaped nest quite high in branches. Lays 2–3 white eggs, sparsely covered with brown and blackish spots. Feeds mainly on fruits, small seeds and insects.

Juvenile male has yellow-orange wash on throat, nape and rump.

Female (right) has whitish supercilium and submoustachial stripe. Adult male (left).

Range *S. zena* is resident on Bahamas, Cuba, Cayman Islands and the island of Cozumel, Mexico.
Status Endemic subspecies *S. z. pretrei* is fairly common resident on Cuba, Isle of Pines and many of the larger cays off north and south coasts.

Red-legged Honeycreeper
Cyanerpes cyaneus 13cm

Local name Aparecido de San Diego.
Taxonomy Polytypic (11).
Description Both sexes have long, slightly decurved bill and red legs. Male breeding plumage is purplish-blue overall with black wings, back and tail; crown metallic turquoise-blue; yellow wing-linings are conspicuous in flight. Non-breeding male similar to female with black wings and tail. Female is olive-green above, with yellowish-white underparts densely streaked with olive.

Voice Song begins with a paired *breep-breep* followed by long series of high-pitched notes *too-too-too-wee-wee-too-too-too*. Call, a repeated, short, high *tseep-tseep*; also, *srelee* and nasal *bizzj*.

Similar species None.

Habitat and behaviour Semi-deciduous and evergreen woodland, tropical karstic woodland. Found mostly in small flocks in mountain and lowland forests. Breeds March–July. Builds cup-shaped nest of grasses, rootlets and other plant fibres. Lays two whitish eggs, with a greenish-blue cast, spotted reddish-brown and lilac at large end. Feeds on nectar, fruits, insects.

Breeding male is purplish blue with black back, wings and tail, and a light-blue crown.

Female is olive-green with streaked underparts.

Range Mexico to southern Brazil.
Status Rare breeding resident on mainland Cuba, but locally common in Sierra del Rosario and Sierra de los Órganos, Pinar del Río province; Santa Clara, Trinidad and high in Sierra Maestra, Holguín, Granma province; also on Cayo Coco.

Cuban Bullfinch

Melopyrrha nigra   14–15cm

Local name Negrito.
Taxonomy Monotypic.
Description Male entirely black with bluish gloss, deep heavy black bill with convex culmen, black iris, white border to closed wing and short, rounded tail. In flight shows short rounded wings with white on primary coverts, inner webs of outer primaries and secondaries; underwing shows white axillaries and underwing-coverts. Female similar but duller, with smaller wing-patch. Juvenile similar to female, with greenish-tipped feathers white on wing reduced.
Voice Thin, prolonged and melodious warble, *ti-ti-tisississiiiitssiiiitsiiii-tooee-toeee*. Call a staccato *chi-dip*; also a thin *tsee*, often repeated.

Similar species None.
Habitat and behaviour Semi-deciduous and evergreen woodland, swamp woodland, tropical karstic woodland, coastal thickets and sandy coast vegetation complex, from sea level to moderate altitudes. Breeds March–August. Pairs have prolonged courtship; males display with wing-flashing that exposes white axillaries. Nest is large and globular, with side entrance, made of dry grasses and leaves, rootlets, hair and feathers. Lays 3–5 whitish eggs with greenish cast, spotted with reddish-brown and lilac concentrated at larger end; females incubate and both adults feed young. Feeds on fruits, seeds, insects.

Adult female is dull black.

Adult male has white band on wing, showing as white primary coverts and underwing-coverts in flight.

Range Cuba.
Status Endemic species; a common breeding resident on Cuba, Isle of Pines and several larger northern cays.

Young male.

Young male.

Male.

Cuban Grassquit

Tiaris canorus   11cm

Local name Tomeguín del Pinar.
Taxonomy Monotypic.
Description Male is dark olive above with grey crown. Mask, throat and large breast-patch are black, outlined by bold yellow-orange band that continues back to a point behind the eye; abdomen grey. Female is similar, with chestnut mask and paler yellow collar; underparts grey. Juvenile similar to female but duller.
Voice Song, a rather harsh *chiri-wechee-wechee, chibiri-wechee*; also, *tsit-tsit-tillio, tsit-tsit-tillio*. Call, a high, often repeated *tsit*.
Similar species Yellow-faced Grassquit has orange or yellow on chin and eyebrow, less olive on back.
Habitat and behaviour Coastal thickets, semi-deciduous and evergreen woodland, pine woodland, swamp woodland, tropical karstic woodland. Breeds April–June. Nest is large and globular with side entrance, and is made of dry grasses, rootlets, hair and other plant fibres; lays 2–3 white eggs, with a greenish-grey cast, spotted with brown and lilac concentrated at large end. Feeds on seeds and small fruits. Social outside breeding season.

Adult female has red-brown face.

Adult male has black face and breast.

Range Cuba.
Status Common endemic in some regions but declining near human settlements. Absent from Isle of Pines and the cays.

Male.

Male.

Female.

Yellow-faced Grassquit

Tiaris olivaceus 11.5cm

Local name Tomeguín de la Tierra.
Taxonomy Polytypic (5).
Description Male has olive upperparts, distinct bright orange supercilium, chin and throat; size of black breast-patch increases with age; small conical bill and grey legs. Female has paler underparts, no black on breast, pale yellow supercilium and chin. Juvenile resembles female with short pale supercilium and chin.
Voice Song a varied series of weak high-pitched trills *zee-zee-zeeeeeee zee-zee-zeeeeeee*; call *quit*.
Similar species Male Cuban Grassquit has large black mask and bold yellow-orange collar; female has chestnut mask and yellow collar.
Habitat and behaviour Coastal thickets, swamp woodland, semi-deciduous and evergreen woodland, second growth, agricultural land, urban areas, from sea level to moderate elevations. Breeds year-round. Nest is low, globular, with side entrance, made of dry grasses, rootlets, hair and other plant fibres at 0.5–6.0m height. Lays 2–4 white eggs with bluish cast, spotted with brown and lilac concentrated at large end. Feeds on seeds, small fruits, tender shoots.

Adult male has bright golden-yellow supercilium and throat, and black breast.

Juvenile resembles female but is duller with pale supercilium and chin.

Adult female has pale yellow supercilium and chin, and lacks black on breast.

Range Resident from Mexico to northern South America and the West Indies, where *T. o. olivaceus* breeds in western Greater Antilles.
Status Very common breeding resident throughout the Cuban archipelago.

Black-faced Grassquit
Tiaris bicolor 11cm

Local name Tomeguín Prieto.
Taxonomy Polytypic (8).
Description Male has dark olive back and wings; sooty, lustreless black head and underparts, vent and undertail-coverts greyish. Female is plain olivaceous-grey above, paler below. Immature duller.
Voice Call note is a soft, musical *tsip*. Song is a buzzing *tik-tseee, tik-tik-tseee*.
Similar species Female Cuban Grassquit has chestnut mask; Yellow-faced Grassquit has pale supercilium and chin.
Habitat and behaviour Usually feeds on ground in coastal thickets. Breeds April–June. Nest is low and globular, with an entrance in the side or bottom; usually lays three whitish eggs, heavily flecked at the broad end. Feeds on seeds and small fruits.

Adult male has black head, throat and breast; blackish to olive abdomen.

Female is brownish-olive overall.

Range West Indies, western and southern Caribbean islands and northern South America. *T. b. bicolor* breeds in Bahamas and Cuba.
Status A small population is known from some of the northern cays.

Zapata Sparrow

Torreornis inexpectata 17cm

Local name Cabrerito de la Ciénaga.
Taxonomy Polytypic (3).
Description Large sparrow with greyish-olive upperparts with dark streaks and chestnut crown with grey median stripe; grey throat with chestnut eye-line, white throat bordered by blackish malar and narrow moustachial stripe; yellow to yellowish breast and abdomen, greyish-olive sides. Black conical bill with pale legs and long broad tail. Juvenile is duller, dark greyish-olive above and light yellow below. Three subspecies are known: Zapata and Cayo Coco subspecies are quite similar with brighter chestnut crown and more yellow on underparts; the easternmost race, from Guantánamo province, is duller, with the crown almost grey. A weak flier.
Voice Song is a high-pitched, buzzy, and cascading *tzi-tzi-tziiiitziiii*. Call is thin *tseep*; also a repeated *pit* and a repeated, metallic *oing*.
Similar species None.
Habitat and behaviour Quite different in the three areas. Zapata marshes are flooded about half the year. On Cayo Coco in evergreen and semi-deciduous woodland, and mangrove woodland. Baitiquirí in coastal thickets. Breeds April–June. Only three cup-shaped nests have been located, all in Zapata swamp, on tussocks surrounded by sawgrass; largest clutch yet discovered comprised two eggs, white with a green wash, spotted brown. In Zapata, during flood season, feeds extensively on eggs of snails (*Pomacea*), and also small lizards; during dry season, feeds on seeds and small invertebrates, often on ground. Eastern birds also take cactus fruits.

Adult, Cayo Coco subspecies.

Adult, Zapata subspecies

Adult, easternmost subspecies from Guantánamo.

Range Cuba.
Status Vulnerable. Rare endemic with populations in three widely separated areas: *T. i. inexpectata* on Zapata Peninsula, specifically on the western section of the marsh; *T. i. varonai* on Cayo Coco and Cayo Romano; and *T. i. sigmani* is found along the coast east of Bahía de Guantánamo, from Tortuguilla to La Magueyera.

Adult, Zapata subspecies.

Adult, Cayo Coco subspecies.

Adult, Zapata subspecies.

Chipping Sparrow

Spizella passerina   14cm

Local name Gorrión de Cabeza Carmelita.
Taxonomy Polytypic (5).
Description Both sexes have fairly long notched tail and whitish wingbars. Winter plumage is brown, narrowly streaked on back with indistinct grey crown-stripe; bill orangish, ear-patch brownish, greyish-buff supercilium, eye-line and lores blackish, rump and underparts grey. Breeding plumage has conspicuous reddish-brown crown, broad white eyebrow, dark bill, eye-line and lores; underparts pale grey. First-year similar to winter adult, but darker.
Voice A dry, sharp *tseep*.
Similar species Clay-coloured Sparrow has whitish median crown-stripe, conspicuous dark-bordered brown ear-patch, pale lores, decidedly pale grey nape and sides of neck, and brown rump.
Habitat and behaviour Bushy areas, coastal thickets, borders of semi-deciduous forest. Feeds on insects and seeds.

Breeding adult has reddish crown and white supercilium. Note blackish eye-line and lores in all plumages.

Range Breeds throughout North America and south to Nicaragua, wintering within this range, also in the Bahamas and Cuba.

Status Very rare winter visitor on Cuba, Cayo Coco and Cayo Guillermo, October–February.

Clay-coloured Sparrow

Spizella pallida   14cm

Local name Gorrión Colorado.
Taxonomy Monotypic.
Description Brown upperparts with black streaks, brown rump appears uniform with back. Striped head, whitish median crown-stripe, broad supercilium and moustache, pale lores, and distinct brown cheek patch bordered by dark brown postocular and moustachial stripes. Grey nape contrasts with buffy-brown and streaked upperparts; faint wingbars. Bill pinkish. Tail fairly long and notched. Non-breeding adults are buffier below. First-year similar to adult, with buff breast and sides.
Voice A thin *tseep*.
Similar species Non-breeding adult and immature Chipping Sparrow have grey rump, reddish crown and dark lores; lack both crown-stripe and dark fore border to cheek-patch; paler and brighter with more pronounced contrast on the face in all plumages.
Habitat and behaviour Bushy savanna, coastal thickets, borders of marshes. Feeds on seeds and insects.

Adult. Note brownish ear-patch bordered with dark brown lines, and pale lores.

Range Breeds in northern North America; winters to Mexico and rarely on Cuba.
Status Rare passage migrant and winter resident on Cuba, Cayo Coco and Cayo Guillermo, October–December and January–February.

Lark Sparrow

Chondestes grammacus 17cm

Note bold chestnut and white stripes on head; tail long with white corners.

Local name Gorrión de Uñas Largas.
Taxonomy Polytypic (2).
Description Brown above with dark streaks; bold head pattern of chestnut crown, white central crown-stripe, wide supercilium, white crescent around eye, reddish ear-coverts with white patch closest to neck, black malar stripe bordered by white moustache and throat. Whitish below, with distinct black spot in centre of breast. Tail long and graduated with white corners.
Voice A sharp *tink* or *jeeet*.
Similar species Adult White-crowned Sparrow has head boldly striped with black and white; tail and breast unmarked.
Habitat and behaviour Bushy savanna, coastal thickets.

Range Breeds in southern Canada, United States and Mexico; majority from Canada and United States winter in southern United States and Mexico.

Status *C. g. grammacus* is very rare passage migrant, October–December and February–March.

Savannah Sparrow

Passerculus sandwichensis 13cm

Adult. Note yellow or buff supraloral area, distinct black malar and moustachial stripes, and heavily streaked breast and sides.

Local name Gorrión de Sabana.
Taxonomy Polytypic (17).
Description Adult brown with buff upperparts; head has pale crown-stripe; yellow or buffy supraloral area and wide whitish supercilium at rear; dark eye-line, narrow malar and moustachial stripes; white throat; breast and sides broadly streaked brown, sometimes with an indistinct central black spot on breast; conical pink bill; notched short tail and pink legs.
Voice Call *chip, chip chip* and *tse*.
Similar species Lincoln's Sparrow has grey supercilium and is more warmly toned and finely streaked below.
Habitat and behaviour Coastal thickets, grassland where it takes seeds and insects.

Range Breeds across North America to central Mexico. Also, far eastern Russia (Chukotka). Winters from southern Canada to central Mexico, Atlantic coast of United States and the West Indies, in northern Bahamas, Cuba and Swan Islands.
Status Uncommon winter visitor, but slightly more common passage migrant on Cuba, Isle of Pines and some northern cays, October–April.

Grasshopper Sparrow

Ammodramus savannarum   12.5cm

Local name Chamberguito.
Taxonomy Polytypic (11).
Description Upperparts brownish-grey, striped with tawny and grey, black centres to wing-coverts, yellow on bend of wing; flat head with whitish median crown-stripe, prominent eye-ring around dark iris in pale buffy face; partial supercilium orange-red in front of eye and buffy behind, dark spot on ear coverts; breast is unmarked cinnamon-buff, abdomen whitish. Large, horn-coloured, longish bill; short tail with diagnostic pointed rectrices and pink legs. Juvenile similar but breast and flanks finely streaked.
Voice Call *kr-it*.
Similar species Savannah Sparrow has heavily streaked breast and black malar stripe. Clay-coloured Sparrow and Chipping Sparrow have buffy or whitish supercilium, grey nape and greyish underparts.

Habitat and behaviour On ground in grasslands, edge of second growth and coastal thickets, usually hidden in long grass. Flies short distances.

Adult has flat striped crown with distinct whitish central stripe, large bill and distinct eye-ring.

Range Breeds from North America through Middle America to Colombia and Ecuador, and the West Indies. Northern subspecies winter on southern parts of breeding range.

Status *A. s. pratensis* is uncommon and easily overlooked winter visitor on Cuba, Isle of Pines and some northern cays, October–May.

Lincoln's Sparrow

Melospiza lincolnii   14cm

Local name Gorrión de Lincoln.
Description Crown has median grey stripe bordered by reddish-brown lateral stripes; eye-ring buffy, wide grey supercilium; note warm buff in submoustachial region. Throat and abdomen are white, breast and sides are warm buff, narrowly streaked black. A distinct dark brown breast-spot may be present. Tail long and rounded.
Voice A hard *chup* and a peculiar buzzy *dzzzzzz*.
Similar species Grasshopper Sparrow is mostly unmarked below and tail feathers are pointed. Savannah Sparrow has more coarsely streaked breast and sides, and lacks warm buffy undertone across breast.
Habitat and behaviour Coastal thickets, grasslands. Furtive and difficult to observe. Feeds on seeds, insects.

Adult. Note grey central crown-stripe, supercilium and ear-coverts, and buffy streaked breast.

Range Breeds in northern and western North America; winters south in United States to Guatemala and the Greater Antilles.

Status Very rare passage migrant and winter resident on Cuba and some northern cays, October–November and January–April.

White-crowned Sparrow
Zonotrichia leucophrys 17cm

Juvenile has duller stripes on head.

Local name Gorrión de Coronilla Blanca.
Taxonomy Polytypic (5).
Description Adult has crown boldly striped black and white, dark streaking on back, narrow white wingbars on brown and buff wings; pointed pinkish or yellowish bill; grey nape, throat, breast and rump, brownish flanks and white undertail-coverts. Immature has broad buff and brown crown-stripes.
Voice A thin *seeet*.
Similar species Lark Sparrow has chestnut and white head stripes and white tail corners. Bobolink is larger, with streaked sides, unmarked wing and sharply pointed tail feathers.
Habitat and behaviour Open forests, grasslands, second growth. Feeds on seeds, insects.

Adult has black-and-white striped head.

Range Breeds in North America; winters in United States, northern Mexico and the Greater Antilles, where it is rare in winter and passage on Cuba and the Bahamas; vagrant in Jamaica.

Status *Z. l. leucophrys* is rare passage migrant on Cuba, Cayo Santa María, Cayo Coco, Cayo Paredón Grande, and a very rare winter resident, October–March.

Rose-breasted Grosbeak

Pheucticus ludovicianus 19–20cm

Local name Degollado.
Taxonomy Monotypic.
Description Very large conical horn-coloured bill. Non-breeding male has brownish-black head and upperparts with white spots on wing-coverts, breast pink and buffy. Male breeding plumage has glossy black head, throat, upperparts and tail, white rump, black wings with two large white patches on coverts, crimson V on central breast, rest of underparts white with streaking on flanks. Female breeding plumage has dark brown striped upperparts, buff central crown-stripe, small whitish wingbars, broad whitish supercilium to behind ear-coverts, white crescent below eye, white malar stripe and throat, breast and flanks buffy-white and heavily streaked. First-winter similar to female but male has pink wash on upper breast, brown edge to blackish upperparts and greyish rump. In flight male has white outer tail feathers, red underwing-coverts with white inner primaries; female has yellow underwing-coverts.
Voice Call short, thin insect-like *chink*.
Similar species None.
Habitat and behaviour Semi-deciduous and evergreen woodland, tropical karstic forest, second growth. Usually in pairs or small flocks. Feeds on fruits and seeds.

Breeding adult male.

First-summer male. Note brownish primaries and whitish edges to head and back feathers.

Adult female.

Range Breeds in North America; winters in Middle America to northern South America and the West Indies, where it is uncommon; regular on passage in the Bahamas, Greater Antilles and Cayman Islands; vagrant in the Lesser Antilles.

Status Uncommon winter visitor and regular passage migrant on Cuba, Isle of Pines and some northern cays, September–May.

Blue Grosbeak

Passerina caerulea 16.5–19.0cm

Local name Azulejón
Taxonomy Polytypic (7).
Description Large heavy conical bill with blackish maxilla and silver mandible, flattened crown, buffy-tan wingbars and long rounded tail. Non-breeding male has feathers edged brownish on body, and whitish undertail-coverts. Breeding male is bright ultramarine-blue, wings dark with rufous-chestnut wingbars (upper is larger in both sexes), black lores and chin. Breeding and non-breeding females have grey-brown or rufous-brown plumage, paler on face and underparts, two dull wingbars, upper more chestnut and lower tawny, blue feathers often visible on rump, wing-coverts and tail. In first-winter both sexes are bright brown with rufous wingbars; male develops blue on head and wings.
Voice Call resonant *click*.
Similar species Indigo Bunting has smaller bill, male is uniformly blue in breeding plumage. Non-breeding male and breeding and non-breeding females have faintly-streaked breast and sides, faint and narrow buff wingbars and whitish throat.
Habitat and behaviour Small flocks usual, occasionally large on migration, feeding on ground or bushes on grass seeds and arthropods in coastal thickets, patches of semi-deciduous and evergreen woodland, pine forest. Constantly flicks and fans tail.

Breeding male is vivid blue with rufouse-chestnut wingbars.

Adult female has warm brown upperparts, paler below.

Range Breeds in United States and Middle America; winters from Middle America to Panama; uncommon on passage in the Bahamas, Cuba, Hispaniola, Puerto Rico and Cayman Islands.

Status *P. c caerulea* is an occasionally common passage migrant and extremely rare winter visitor on Cuba, Isle of Pines, many northern cays and Cayo Cantiles off the south coast, August–May.

Indigo Bunting

Passerina cyanea 14cm

Local name Azulejo.
Taxonomy Monotypic.
Description Small conical bill and short tail. Non-breeding male appears brown with blue feathers partially covered by brownish edges on head, wings and breast. Non-breeding male and breeding female are cinnamon-brown overall with white throat, faint buffy wingbars and light streaking on breast and sides; some have blue tinge on wing-coverts, edge of wing, rump and tail. Breeding male is entirely blue, becoming indigo on head and breast; in flight shows brownish-grey flight and tail feathers. First-year male is brown with blue feathers emerging on body, head, wings and tail.
Voice Males sing on late spring migration, long series of insect-like *swee swee sweer* and short trills or clicks; call *chip*.
Similar species Blue Grosbeak is larger with larger bill; male is deep blue with rufous-chestnut wingbars.
Habitat and behaviour Second growth and semi-deciduous and evergreen woodlands, coastal thickets, citrus plantations, areas with abundant *Acacia*, pines, open woodland, meadows and sandy coast vegetation complex. Usually small flocks.

Breeding adult male is indigo-blue, darker on head and breast.

First-year male developing adult plumage in spring.

Breeding female. Note buffy wingbars and faintly-streaked breast and sides, and blue on wing-coverts and tail.

Range Breeds in North America; winters from Middle America to Panama and the West Indies, in the Bahamas, Greater Antilles and Cayman Islands; vagrant in the Lesser Antilles.

Status Common winter visitor and passage migrant throughout the Cuban archipelago, September–May.

Painted Bunting

Passerina ciris 14cm

Adult female has bright-green upperparts and yellowish-green underparts. First-winter male resembles female.

Local name Mariposa.
Taxonomy Polytypic (2).
Description Male is unmistakable with purple-blue head, yellow-green mantle, red rump, throat, underparts and eye-ring. Slightly duller in winter. Small conical bill. Female and first-winter male have faint yellowish eye-ring, olive-green upperparts, brighter on back than rump, and olive-yellow underparts, greyer on breast and brighter yellow on abdomen.
Voice Silent.
Similar species None.
Habitat and behaviour Coastal thickets, semi-deciduous woodland.

Adult male has blue head, red underparts and yellowish-green back.

Range Breeds in south-eastern United States; winters in the Bahamas and Cuba, where now fairly common; vagrant on Jamaica.

Status *P. c. ciris* is fairly common passage migrant and winter visitor on Cuba, Isle of Pines and northern cays, October–April.

Dickcissel

Spiza americana 15–18cm

Local name Gorrión de Pecho Amarillo.
Taxonomy Monotypic.
Description Adult has brownish-grey upperparts with black streaking, rufous coverts (shoulder), dark wings edged with buff, greyish crown and ear-coverts, long yellowish supercilium, black malar stripe and broad pale submoustachial stripe, white chin and throat, grey conical bill, and pointed central tail feathers; non-breeding male has pale but defined breast-patch. Breeding male has grey nape, black V on throat and central breast, yellow breast, rest of underparts greyish-yellow. Breeding female smaller with yellowish wash on breast (black absent) and blurred streaking on sides and flanks. First-winter is dull greyish-brown overall, rufous coverts absent, back heavily streaked and breast faintly streaked.
Voice Silent.
Similar species None.
Habitat and behaviour Coastal thickets. Feeds on seeds.

Adult female lacks black bib, and has smaller chestnut wing-patch and breast washed yellow.

Adult male in breeding plumage.

Range Breeds from southern Canada over the central and eastern United States; winters in Middle America and especially in northern South America.

Status Rare passage migrant, mostly in western Cuba, September–December and February–May.

Bobolink

Dolichonyx oryzivorus 18.5cm

Local name Chambergo.
Taxonomy Monotypic.
Description Non-breeding adults have black crown with buff nape and central crown-stripe, black postocular stripe, buffy throat, bright yellow-buff plumage and rump, black stripes on back, two faint wingbars, black streaking on sides, flanks and lower abdomen, heavy pinkish bill and tail with sharply pointed feathers. Male in early spring shows buffy edges to feathers. Breeding male is black, except pale yellowish-buff hindcrown and nape, whitish-grey rump, white scapulars and gold streaks on back, wing feathers edged whitish-buff, dark bill. Breeding female similar to non-breeding adult with greyish-buff nape, whitish throat, fainter streaking on flanks of buffy-grey underparts and grey rump.
Voice Loud *pink* repeated.
Similar species All sparrows are smaller and lack streaking on sides and lower abdomen, and pointed tail feathers.
Habitat and behaviour Males and females often in separate small flocks foraging for seeds and insects in rough pasture, grassland, littoral shrubland and herbaceous wetlands.

Non-breeding adult female. Tail feathers are sharply pointed in all plumages.

Breeding male has black head and creamy hindcrown and nape.

Range Breeds in North America; winters in central South America; on passage through the West Indies where it is most frequent in the Bahamas, Cuba, Jamaica, Cayman Islands, Providencia and San Andrés.
Status Uncommon but regular passage migrant throughout the Cuban archipelago, August–December and March–June.

Red-shouldered Blackbird
Agelaius assimilis 21cm

Local name Mayito de Ciénaga.
Taxonomy Monotypic.
Description Both sexes black and slightly glossy; bill sharply pointed, wide at base. Male has orange-red patch on shoulder edged buffy-yellowish, only entirely visible in flight or when displaying. Female is entirely black. First-year male similar to adult but duller, with brownish patch on shoulder. Both sexes have squarish-ended tail with slightly pointed feather tips.
Voice Both sexes sing. Frequently repeated loud harsh creaks, *o-weeheeee-o weeheeee*; also, a short *chek-chek*.
Similar species Tawny-shouldered Blackbird is smaller, has orange-buff shoulder patch, smaller bill and notched, rounded tail. Cuban Blackbird is larger and entirely glossy, with longer, thicker decurved bill. Male Shiny Cowbird is entirely black with greenish gloss on wings.
Habitat and behaviour Swamps, marshes, swamp forest edge. Feeds on insects, seeds, nectar. Very social, forming flocks with Cuban Blackbird and Tawny-shouldered Blackbird outside breeding season. Breeds May–August. Builds a deep, cup-shaped nest among tall sawgrass and reeds and lays 3–4 pale bluish-white eggs, heavily spotted and scrawled with reddish-brown and pale purple. Monogamous.

Adult male has red shoulder patch, edged yellow to buff.

Adult female is entirely black. Both sexes sing.

Immature male has partial tawny-brown epaulette.

Range Cuba.
Status Endemic; common but extremely local resident at Laguna de Lugones, Guanahacabibes Peninsula and Laguna Jovero (south of Guane) in Pinar del Río province; Lagunita, Salinas de Bidos, Itabo and Ciénaga de Zapata. Rare on Isle of Pines at Ciénaga de Lanier.

Adult male.

Adult female.

Adult female.

Tawny-shouldered Blackbird

Agelaius humeralis 19–22cm

Local name Mayito.
Taxonomy Polytypic (2).
Description Black; both sexes with orange-buff shoulder patch, smaller and duller in female, regularly visible in flight or when displaying. Tail slightly notched, rounded at tip. Juvenile is dull black, with smaller and paler shoulder patch, sometimes visible only in flight.
Voice Short whistles, harsh creaks, *weee-weee*, and a nasal *enk*. When giving strong *cheek* call, often repeated, invariably cocks the tail.
Similar species Male Red-shouldered Blackbird is larger and has orange-red shoulder patch; female is entirely black with larger bill; both sexes have slightly pointed tail. Cuban Blackbird is larger and entirely black with longer, thicker decurved bill and squarish tail. Male Shiny Cowbird is entirely black with greenish gloss on wings.
Habitat and behaviour Forest edge, semi-deciduous and evergreen woodland, swamp woodland, tropical karstic woodland, wetlands, rainforest, agricultural fields. Highly social. Breeds April–August. Builds cup-shaped nest in a tree, of grasses, hair, and occasionally some feathers; lays 3–4 greenish-white, brown-spotted eggs. Feeds on bees, seeds, nectar, fruits, small lizards.

Adult male. Both males and females have orange-buff shoulder patch.

Adult female has smaller shoulder patches than male.

Range Cuba and western Haiti.
Status Subspecies *A. h. humeralis* is common breeding resident on Cuba and several larger cays on both coasts. Not known from the Isle of Pines. Endemic subspecies *A. h. scopulus* is resident on Cayo Cantiles.

Eastern Meadowlark

Sturnella magna 23cm

Immature.

Local name Sabanero.
Taxonomy Polytypic (16).
Description Heavy-bodied with long bill; short tail has white outer rectrices. Breeding adult has crown and upperparts heavily marked with brown, black, rufous and whitish, and may form stripes on back, wide pale supercilium, bright yellow throat and underparts with broad black breast-band and darkly streaked sides and flanks, pale legs. Non-breeding adult has brown streaking on crown and upperparts and less distinct breast-band. Juvenile similar to adult but paler, with greyish throat. In flight, rapid shallow wingbeats alternate with short glides; tail fanned.
Voice A sweet, slurred whistle.
Similar species None.
Habitat and behaviour Mainly terrestrial in grassland and wetlands. Breeds April–July. Builds domed nest among grasses; lays 4–5 white eggs with reddish-brown spots concentrated towards the large end. Feeds on insects, small lizards, worms, seeds.

Adult has black V on breast when breeding; black V on breast is scaled with yellow in non-breeding birds. This species is yellow below in all plumages.

Range Breeds in United States, Middle America, South America and the West Indies, on Cuba (populations outside North America are sedentary); winters in United States south throughout the breeding range.
Status *S. m. hippocrepis* is an endemic subspecies (may warrant endemic species status) on Cuba, Isle of Pines and several northern cays.

Cuban Blackbird

Ptiloxena atroviolacea 27cm

Local name Totí.
Taxonomy Monotypic.
Description Entirely black, sexes similar with violaceous metallic sheen; thick decurved bill, brown eye and flat square-ended tail. Female is slightly smaller. Juvenile is duller.
Voice A highly variable *tew-tew-tew*, or *tee-leeoo* often repeated, a prolonged, nasal *enk*, a chatter call, and a simple *kik*, often repeated.
Similar species Adult Greater Antillean Grackle has yellow eye, longer thinner bill, and V- or keel-shaped tail. Female Red-shouldered Blackbird is smaller, with slightly pointed tips to tail feathers. Male Shiny Cowbird is smaller with greenish gloss on wings.
Habitat and behaviour Semi-deciduous and evergreen woodland, swamp woodland, tropical karstic woodland, riparian woodland, second growth, urban areas. Breeds March–July. Cup-shaped nest is built on base of palm fronds, among palm seed clusters or bromeliads, and made of dry grasses, rootlets, hair and feathers. Lays 3–4 greyish-white eggs with grey and brown spots concentrated at large end. Omnivorous.

Adult.

Juvenile plumage is dull black.

Range Cuba.
Status Common endemic, widespread across Cuba and some northern cays. Absent from Isle of Pines.

Adult has black plumage with violaceus iridescence, brown iris and square-ended tail.

Adult.

Greater Antillean Grackle

Quiscalus niger 25–30cm

Local name Hachuela, Chichinguaco.
Taxonomy Polytypic (7).
Description Adults of both subspecies have yellow iris and long, pointed bill. Male is large with glossy metallic violet-blue to blackish-violet body, wings with iridescent bronze-green and long V- or keel-shaped bluish-green tail. Female duller, tail smaller with slight V. Juvenile dull brownish-black, tail normal shape, iris light brown.
Voice Call *ching ching ching;* musical four-syllable fluting song; also, sharp *cluck*, and raucous begging call of juvenile.
Similar species Female Red-shouldered Blackbird is smaller, with conical bill, and slightly pointed tips to tail feathers.
Habitat and behaviour Semi-deciduous and evergreen woodland, tropical karstic woodland, swamp woodland, coastal thickets, mangroves, urban areas. Breeds March–July. Cup-shaped nest is built by both adults of grasses and mud, usually on base of palm fronds or large bromeliads. Lays 3–5 olivaceous eggs with reddish-brown spots and scrawls. Gregarious, in all habitats. Omnivorous. Both subspecies have large roosting colonies; prolonged courtship.

Adult female is smaller than male and V-shape of tail is reduced.

Juvenile is dull, mottled black and brown with dark iris.

Adult male is glossy violet-blue with long, deep V-shaped tail.

Range Greater Antillean endemic.
Status Endemic subspecies *Q. n. gundlachii* is abundant throughout the Cuban archipelago and northern cays. *Q. n. caribaeus* is fairly common in western Cuba, Isle of Pines and southern cays.

Shiny Cowbird

Molothrus bonariensis 18–20cm

Juvenile has faint streaking on underparts.

Local name Pájaro Vaquero.

Taxonomy Polytypic (7).

Description All plumages have dark iris and short, sharply pointed, conical bill. Male is glossy purple-black and greenish gloss on wings. Female has dull greyish-brown plumage, pale buff supercilium and underparts. Juvenile similar to female except supercilium is more indistinct, faint buffy wingbars and underparts softly streaked brownish-grey.

Voice Song series of high warbles and a trill. Call rolling *chuck*.

Similar species Tawny-shouldered and Red-shouldered Blackbirds have shoulder patches. Female Red-shouldered Blackbird lacks glossy look, and has larger bill and more pointed tail feathers. Cuban Blackbird is larger with longer, thick decurved bill and larger tail.

Habitat and behaviour Second growth, rural areas, swamp woodland. Brood parasite; in the region lays eggs in nests of mockingbirds, kingbirds, endemic vireos and warblers, and icterids; host species raises the cowbird young while young of host are often starved or ejected. Feeds on grain, hence presence on farms.

Adult male is glossy purple-black with greenish gloss on wings.

Range South America, Panama and Costa Rica. It is established in Florida and the West Indies, where it colonised Puerto Rico in the 1940s, Cuba in the 1970s, Jamaica in 1993 and the Bahamas in 1994.

Status Resident *M. b. minimus* is common in parts of Cuba, Isle of Pines and some northern cays.

Cuban Oriole
Icterus melanopsis 20cm

Local name Solibio.
Taxonomy Monotypic.
Description Sexes alike. Black with yellow patches on shoulder, underwing-coverts, rump and tail-coverts. Bill slightly decurved with greyish base. Long and slightly rounded tail. Juvenile is duller, olive-yellow overall with just a few black markings on face. Immature is olivaceous-green with black forehead, throat and breast; black wings with yellowish-green shoulder patches.
Voice Song, a series of 8–12 clear, varied, down-slurred whistled notes. Call, a harsh *tick*.
Similar species First-spring male Orchard Oriole has black markings on back, contrasting with unmarked rump; female has two greenish-white wingbars.
Habitat and behaviour Semi-deciduous and evergreen woodland, swamp woodland, tropical karstic woodland, second growth, urban areas. Flight is undulating, producing churring sound. Breeds February–July. Builds elaborate globular nest from palm fibres sewn to underside of palm frond, with side entrance. Lays three greenish-white eggs, with brown and lilac spots concentrated at large end. Found from sea level to at least 1,300m in eastern third of Cuba. Feeds on fruits, nectar, butterflies and other insects.

Immature is greenish-yellow with black throat.

Adult is black overall, apart from yellow shoulder, rump, undertail-coverts and thighs.

Range Endemic to Cuba.
Status Common on Cuba, Isle of Pines and most northern cays.

Adult.

Immature.

Immature.

Adult.

Orchard Oriole

Icterus spurius   18cm

Adult female has yellowish-green underparts and narrow white wingbars.

Local name Turpial de Huertos.
Taxonomy Polytypic (2).
Description Male has black head, breast and upper back; chestnut abdomen, lower back, rump and lesser wing-coverts. Female has olive upperparts, bright yellowish-green underparts; wing dark with two narrow whitish wingbars. Both sexes have slightly rounded tails. Immature like female; first-year male has black on chin, throat and chest, and back mottled with black.
Voice A soft *chuck*.
Similar species Baltimore Oriole is slightly larger, underparts orange-yellow, bill longer. Cuban Oriole (immature and subadult) has plain wings, olivaceous back and yellowish-green patches on shoulder, lower back, rump and undertail-coverts.
Habitat and behaviour Semi-deciduous and evergreen woodland, tropical karstic woodland and coastal thickets. Usually solitary. Feeds on insects, spiders, worms and fruits.

Adult male has black head, throat, upper back and tail, and chestnut rump, upperparts and patches on coverts.

Range Breeds in southern Canada, eastern United States and Mexico; winters to Colombia and Venezuela.
Status *I. s. spurius* is rare passage migrant on Cuba and some northern cays; very rare during autumn, less so in spring, October–November and March–May. One winter record (January).

Baltimore Oriole

Icterus galbula 18–21cm

Local name Turpial.
Taxonomy Monotypic.
Description Male has black hood and back; rest of underparts orange; orange bar on shoulder and one white wingbar; orange lower back, rump, uppertail-coverts. Tail with outer retrices broadly orange, with a narrow black band, and black central feathers. Grey bill is sharply pointed. In flight males show black head and distinct orange underparts; wing-linings and tail are mostly orange, tail is black-based with black central retrices. Adult female has crown and upper back mostly brownish-orange, may have blackish feathers on head, two white wingbars, whitish throat, underparts dull orange-yellow. Immature male similar to adult female, but with black feathers emerging around face and throat; immature female duller.
Voice Usually silent, occasionally a chattering threat call.
Similar species Immature male Orchard Oriole is yellowish-green below, with shorter bill. Hooded Oriole (vagrant) is also yellowish-green below, but bill longer and decidedly decurved; female has pale yellow underparts; first-year male has pale lores, black on chin and throat.
Habitat and behaviour Semi-deciduous and evergreen woodlands, tropical karstic woodland, swamp woodland, second growth. Usually solitary.

Adult female.

Adult male.

Range Breeds in North America; winters in south-eastern North America, Middle America to northern South America and the West Indies, where it is rare in winter but regular on passage, in the Bahamas, Cuba, Jamaica and Cayman Islands.

Status Fairly common passage migrant and rare winter visitor on Cuba, the Isle of Pines, several northern cays and Cayo Anclitas off south coast, September–May.

APPENDICES

Appendix A
Vagrants recorded on the Cuban archipelago

Common Loon (Great Northern Diver)
Gavia immer   81cm
Local name Somormujo.
Taxonomy Monotypic.
Description Non-breeding plumage dark grey above, white below. Forehead steep; bill straight, stout, and greyish.
Range Breeds in northern half of North America, Greenland and Iceland. Winters along both Pacific and Atlantic coasts of North America south to Mexico; in the Old World, south to Mediterranean and north-west Africa.
Status Five records between 1971 and 1986.

Great Shearwater
Ardenna gravis   46cm
Local name Pampero Grande.
Taxonomy Monotypic.
Description Dark brown cap contrasts with white cheek and collar (sometimes incomplete) and greyish-brown back; dark bill; and white tipped uppertail-coverts. In flight wings long and narrow, white underwing with conspicuous dark tailing edge.
Range South Atlantic, the Falkland Islands and the south-west Indian Ocean. Pelagic during the non-breeding season, migrating to the North Atlantic.
Status One record, in 2015.

Cory's Shearwater
Calonectris diomedea   51cm
Local name Pampero de Cory.
Taxonomy Monotypic.
Description Upperparts greyish-brown; underparts white; thick yellow bill with dark tip.
Range Open-water Atlantic Ocean wanderer; reaches Mediterranean Sea and Indian Ocean. Breeds in Azores, Madeira and Canaries and on Mediterranean coast.
Status Six records between 1951 and 2016.

Sooty Shearwater
Ardenna grisea 43cm
Local name Pájaro de las Tempestades.
Taxonomy Monotypic.
Description Dark brown overall, except whitish underwing-coverts; bill black. Arcing flight.
Range Breeds at southern latitudes: New Zealand, islands near southern tip of South America, Falkland Islands.
Status Nine records between 1936 and 2007.

Wilson's Storm-Petrel
Oceanites oceanicus 18cm
Local name Pamperito de Wilson.
Taxonomy Polytypic (2).
Description Overall dark. In flight feet show beyond tip of tail, inner wing shows pale diagonal bar, conspicuous white band on rump and white patch on lower flank. Legs long; toes with yellow webs, although web colour is very inconspicuous; tail short and square, wings rounded. Hovers over surface of water while feeding, pattering the surface with brown feet extended.
Range *O. o. oceanicus* of Southern Hemisphere is also common in North Atlantic in non-breeding season. Breeds in Antarctica, islands surrounding southern tip of South America and southern Indian Ocean.
Status Seven records between 1946 and 1983.

Leach's Storm-Petrel
Hydrobates leucorhous 20cm
Local name Pamperito de Tempestades.
Taxonomy Polytypic (4).
Description Dark brown, with divided white band on rump; wings long, pointed and angled with pale diagonal bar on inner wing; bill, legs and feet (including webs) black; tail forked.
Range *O. l. leucorhoa* is cold-water oceanic species of Northern Hemisphere, migrating to tropical waters in northern winter. Breeds on offshore islands.
Status Five records between 1934 and 1975.

Band-rumped Storm-Petrel
Hydrobates castro 20cm
Local name Pamperito de Castro.
Taxonomy Monotypic.
Description Very dark brown, with pale inner wingbar and conspicuous broad white rump-patch; bill, legs and feet (including webs) black; tail squarish or slightly forked.
Range Oceanic bird of both tropical Atlantic and Pacific Oceans. Breeds on Salvage, Madeira, Cape Verde, Ascension and St Helena islands in the Atlantic, and also Galápagos and Cocos islands in the Pacific.
Status Four records between 1964 and 2012.

Red-billed Tropicbird
Phaethon aethereus 48cm, excluding tail-streamers
Local name Rabijunco de Pico Rojo.
Taxonomy Polytypic (3).
Description Much like a tern with very long tail. White, with black bars on back and wingtips conspicuously black. Bill is robust, red. Immature has fine black bars above, but white crown and black collar.
Range *P. a. mesonauta* breeds on islands in eastern Pacific, eastern Caribbean Sea (Vieques island to Little Tobago, Isla Blanquilla to Isla Aves) and eastern Atlantic Ocean.
Status Six records between 1947 and 1988.

Masked Booby
Sula dactylatra 81cm
Local name Pájaro Bobo de Cara Azul.
Taxonomy Polytypic (4).
Description White except black tail, upperwing flight feathers and naked skin around base of bill. Bill and legs orange-yellow to olive. Immature has brown upperparts, white rump and collar on upper back, white underparts, and grey legs. In flight shows distinct white wing-linings.
Range *S. d. dactylatra* breeds in southern Bahamas, south-west Jamaica, Virgin Islands, islands off north-eastern Brazil, Ascension Island.
Status Ten records between 1948 and 2016.

Red-footed Booby
Sula sula 66–76cm
Local name Pájaro Bobo Blanco.
Taxonomy Polytypic (3).
Description Two colour morphs, brown and white. White-morph birds are white with black primaries and secondaries. Brown-morph birds are brown with contrasting white tail, rump and undertail-coverts. Both morphs have bluish bill, black band on posterior edge of wings and red legs. Breeding birds have bare skin around eye blue, and pink around base of bill. In flight shows black bar on underwing median primary coverts. Immature is brown all over with blackish bill and olive to yellow legs. Brown morph has never been observed on Cuba.
Range *S. s. sula* breeds on cays in the southern Caribbean, Belize and the West Indies; tropical south-west Atlantic islands.
Status Nine records between 1870 and 2015.

Northern Gannet
Morus bassanus 94cm
Local name Pájaro Bobo del Norte.
Taxonomy Monotypic.
Description White with mustard-coloured wash on head and neck, black wingtips and legs; grey bill. Juvenile mostly dusky with white belly.
Range Breeds on North Atlantic islands. American populations winter south to Gulf of Mexico.
Status Five records between 1993 and 2017.

Scarlet Ibis
Eudocimus ruber 58cm
Local name Coco Rojo.
Taxonomy Monotypic.
Description Brilliant scarlet with black wingtips. Long decurved bill.
Range South America, from Venezuela to Brazil, mostly near coast.
Status Eleven records between 19th century and 1993.

White-faced Ibis
Plegadis chichi 58cm
Local name Coco Prieto de Cara Blanca.
Taxonomy Monotypic.
Description Red facial skin surrounded by white feathers, long decurved grey bill with long reddish legs. Reddish-brown head, neck, upper back and underparts; lower back, wings and tail glossy iridescent bronze-green. Non-breeding plumage has head and neck finely streaked white, the rest of body iridescent bronze-green. Juvenile duller than adult in winter.
Range Western North America, wintering south to northern Central America. Also breeds in southern and central South America.
Status One record, in 2018.

White-faced Whistling-Duck
Dendrocygna viduata 43cm
Local name Yaguasa Cariblanca.
Taxonomy Monotypic.
Description Distinct white face, back of head and neck black, brown back with buff feather edges, dark brown wings and tail. Underparts with reddish-chestnut breast, central belly black, sides finely barred in black and white. Black bill; grey legs. Juvenile has entirely grey head and neck.
Range Breeds in Costa Rica through most of South America, Africa, Madagascar and the Comoros Islands.
Status Four records between 1858 and 1947.

Greater White-fronted Goose
Anser albifrons 73cm
Local name Guanana.
Taxonomy Polytypic (4).
Description Greyish-brown upperparts, underparts with irregular dark barring. Pink bill, with white around base. Tail-coverts white, legs yellow to orange. First-winter similar to adult but without barring on underparts.
Range *A. a. gambelli* breeds in northern North America, winters to southern United States and north-east Mexico.
Status Three records between 19th century and 2012.

Snow Goose
Anser caerulescens 71cm
Local name Guanana Prieta.
Taxonomy Polytypic (2).
Description Two colour morphs, white and dark; adults of both have pink bill and legs. White morph has black primaries and rust-coloured feathers on face. Dark morph is greyish-brown with white head and neck, black primaries and secondaries. Some may show dark brown breast with white belly, while in others the dark brown covers most of underparts and white is restricted to lower abdomen. Juvenile white morph has grey back with dark legs and bill; dark morph is almost entirely greyish-brown, including bill and legs.
Range *A. c. caerulescens* breeds in the Arctic; winters to southern United States and northern Mexico.
Status Uncommon winter visitor in 19th century. Three records between 1925 and 1985.

Canada Goose
Branta canadensis 64–110cm
Local name Ganso del Canadá.
Taxonomy Polytypic (7).
Description Head and long neck black with wide chin strap, back and wings barred brownish-grey, pale breast, brownish-grey underparts darker on sides, dark terminal band on white tail.
Range Breeds in northern North America, wintering south to Mexico and northern Florida (unknown subspecies).
Status Three records between 1966 and 2013.

Tundra Swan
Cygnus columbianus 132cm
Local name Cisne.
Taxonomy Polytypic (2).
Description White, with long neck and black bill and legs. Most birds have yellow spot in front of eye. Juvenile is pale ashy-grey with pink bill, dusky at tip.
Range *C. c. columbianus* breeds in northern North America and north-eastern Siberia, wintering in western and eastern coastal United States.
Status Two records, in 1944 and 1990.

Mottled Duck
Anas fulvigula   56cm
Local name None.
Taxonomy Polytypic (2).
Description Dark brown. In flight white underwing contrasts with dark brown body. Speculum violet, bordered with black lines but may show white on trailing edge. Bill olive-yellow, with black spot near bill on lower face. Legs orange; tail has no white.
Range *A. f. fulvigula* breeds in southern and eastern United States.
Status One record, in 2009.

Cinnamon Teal
Spatula cyanoptera   41cm
Local name Pato Canelo.
Taxonomy Monotypic.
Description Male is cinnamon-red with dark brown mottling on back. In flight, body appears very dark overall. Female is ashy-brown, mottled and variegated. Both sexes have green iridescent speculum with black inner and outer borders and blue patch on forewing.
Range Breeds in western North America and northern Mexico. Also, resident in South America, wintering south to northern South America.
Status Ten records between 1917 and 2017.

Surf Scoter
Melanitta perspicillata   51cm
Local name None.
Taxonomy Monotypic.
Description Male all black except for white patches on nape and forehead; massive distinctly patterned bill with red, orange, white and black. First-winter male is pale brown with darkish area at side of face; bill similar to adult male but paler, and red may show by December. Female all brown with distinct dark cap, white nape and white spots on side of face; dark grey bill with vertical white patch at base.
Range Breeds in northern North America from western Alaska to Labrador, and winters from Aleutian Islands south to Baja California and Pacific coast of Mexico, and along Atlantic coast as far as southern Florida and the Gulf coast.
Status One record in 2015, the first for the West Indies.

Bufflehead
Bucephala albeola   36cm
Local name Pato Moñudo.
Taxonomy Monotypic.
Description Male's head is black with green or purple gloss and large white patch, black back and white underparts. First-winter has smaller head patches. Female is dark brown, with white breast and belly, and small white patch on sides of head. Both sexes have very short bluish bill and broad white wingbar on inner wing, very distinct in flight. Juvenile resembles adult female.
Range Breeds in Canada and north-western United States, wintering to southern United States and Mexico.
Status Four records between 1857 and 2017.

Swainson's Hawk
Buteo swainsoni   117–137cm
Local name Gavilán de Swainson.
Taxonomy Monotypic.
Description Light morph has brown upperparts with broad brown breast-band and contrasting white throat, lores and forehead; white abdomen mostly streaked on side with a few spots on central abdomen. Dark morph is dark overall with whitish undertail-coverts. Intermediate morphs have all-barred underparts. Tail grey, faintly banded with dark broad subterminal band. Juvenile has buffy feathers edges. In flight long narrow wings tapered at tip, with a contrasting white wing lining.
Range Western North America from Alaska and north-west Canada to Northern Mexico; winters in south-central South America, mostly on the pampas of Northern Argentina.
Status Several records between 2010 and 2017; possibly a rare passage migrant.

Virginia Rail
Rallus limicola   24cm
Local name Gallinuela de Virginia.
Taxonomy Polytypic (4).
Description Brown, mottled above, with flanks strongly barred white; wings and breast rusty-brown, cheeks grey. Legs and bill long, usually reddish. Juvenile very dark, with mostly black underparts.
Range *R. l. limicola* breeds in North America and winters to Mexico.
Status Four records between 19th century and 1995.

Long-billed Curlew
Numenius americanus 58cm
Local name Zarapico Pico Cimitarra Grande.
Taxonomy Monotypic.
Description Large, with extremely long, thin decurved bill; bluish-grey legs; cinnamon-brown above, buffy below. In flight shows bright cinnamon underwing. Usually solitary. Juvenile has shorter bill.
Range Breeds in western and central North America, wintering south to Costa Rica.
Status Six records between 1848 and 2005.

Hudsonian Godwit
Limosa haemastica 38cm
Local name Avoceta Pechirrojo.
Taxonomy Monotypic.
Description Large shorebird. Bill very long, bicoloured, and slightly upturned. Non-breeding is grey-brown above, with greyish-brown wash on neck and upper breast. Breeding plumage blackish and mottled above, reddish-cinnamon below. In flight black wing-lining conspicuous, white median stripe on upperwing and tail has broad black subterminal band, tipped whitish. Juvenile resembles breeding adult, but paler, with scaly brownish-black back and buffy underparts.
Range Breeds in northern North America, wintering in southern South America.
Status Four records between 19th century and 2004.

Marbled Godwit
Limosa fedoa 40–51cm
Local name Avoceta Parda.
Taxonomy Polytypic (2).
Description Large, with a very long, slightly upturned, bicoloured bill (pink base with dark tip section). Buffy-brown, mottled above; in flight shows cinnamon underwings. Non-breeding adult is pale brown below with little barring. Breeding underparts are heavily barred reddish-brown. Juvenile resembles winter adult, with cinnamon-buff underparts and a few bars on flanks.
Range Breeds in central North America, wintering south along both coasts to Colombia, rarely to northern Chile.
Status Nine records between 19th century and 2019.

Buff-breasted Sandpiper
Calidris subruficollis 19–22cm
Local name None.
Taxonomy Monotypic.
Description Resembles a plover. Slender, with small round head, dark crown and dark eye, face and underparts warm buff with spots on sides of neck and breast; upperparts with dark brown-centred feathers with pale edge; pointed black bill and yellow legs.
Range Breeds in Arctic North America north of 60°N; winters in southern South America. Rare on passage in the West Indies.
Status Two records, in the 19th century and 2017.

Ruff
Calidris pugnax 25–35cm
Local name None.
Taxonomy Monotypic.
Description Large shorebird with small head; bill dark or orange-pink, as long as the head and drooped at tip. Non-breeding plumage greyish-brown above with pale fringes, giving a scaly look. Below, mostly white and may have spotting or streaking on sides. Legs orange-yellow. Breeding male has variable ear-tufts and neck ruff in chestnut, bare face with wattles and warts that can be orange. Female brownish, with pale underparts. Wing with dark-centred feathers, may also be barred in black and reddish-brown with pale margins.
Range Breeds from northern Europe east to Sea of Okhotsk, and winters from western Europe, Mediterranean and sub-Saharan Africa through Middle East to South-East Asia.
Status One record, in 2017.

Wilson's Phalarope
Phalaropus tricolor 23cm
Local name None.
Taxonomy Monotypic.
Description Non-breeding adult has grey upperparts with white uppertail-coverts; white below. Crown grey, lores and eye-stripe indistinctly grey, cheeks practically unmarked, wings entirely grey, bill long and notably thin, and legs yellow. In breeding plumage, legs are black; female has rich cinnamon stripe along neck and back. Male is duller. Juvenile resembles winter plumage adults, but is browner above with buffy breast. Often swims, but seen on land much more frequently than other phalaropes.
Range. Central and western North America, wintering in

South America from Peru and Bolivia south to Argentina.
Status Five records between 1950s and 2018.

Red-necked Phalarope
Phalaropus lobatus   18cm
Local name Zarapico Nadador Rojo.
Taxonomy Monotypic.
Description Smallest phalarope with needle-like bill. Non-breeding adult grey above with blackish nape and whitish stripes on back; white below with a black patch through eye; white stripe on wing. Thin bill shorter than head. In breeding plumage, both sexes have buffy stripes on back and reddish-brown patches on sides of neck. Male is duller. Juvenile resembles breeding adult, but duller. Almost always on water.
Range Breeds in northern North America, Greenland and northern Eurasia. American populations winter in South America.
Status Four records between 1953 and 2016.

Red Phalarope (Grey Phalarope)
Phalaropus fulicarius   22cm
Local name Zarapico Nadador.
Taxonomy Monotypic.
Description Winter adult pale grey above with black nape, black patch through eye. Bill black, slightly thick and shorter than head (sometimes with yellow base); white below, and white stripe on wing. Breeding bird has bill yellowish with dark tip, cheeks white, underparts uniformly reddish-brown. Male is paler. Juvenile similar to breeding male but paler. Almost always on water.
Range Breeds in northern North America, in the Nearctic from Greenland and Iceland east through Arctic islands to northern Siberia. North American populations winter south to South America.
Status Four records between 1963 and 2018.

South Polar Skua
Stercorarius maccormicki   58cm
Local name Eskúa del Polo Sur.
Taxonomy Monotypic.
Description Dark brown, except very bold white patch at base of primaries; densely streaked upperparts and underparts. Wings very broad. Juvenile has less streaked underparts. Flight powerful, and habits piratical, pursuing other seabirds.
Range Breeds on South Shetland Islands and Antarctica; ranges at sea regularly to North Pacific and North Atlantic.
Status One record, in 1986.

Long-tailed Jaeger (Long-tailed Skua)
Stercorarius longicaudus   50–58cm
Local name Estercorario Rabero.
Taxonomy Polytypic (2).
Description Slender, with long pointed wings, a long tail and a short thick bill. Adult light morph has black cap, yellow wash on sides of neck and nape, and white breast darkening to grey towards tail. In flight grey above, with contrasting darker flight feathers, dark underwings except base of outer primaries which are white. Rare dark morph is entirely dark brown below. Central tail feathers are very long and sharply pointed. Winter adult has distinct barring on undertail-coverts and flanks, like juvenile, grey collar and indistinct crown patch, and shorter central tail feathers. Juveniles of both colour morphs have short central tail feathers, strong barring on underwing and undertail-coverts.
Range Breeds in Arctic and subarctic uplands from Scandinavia to Siberia, Greenland and North America; winters in sub-Antarctic and off southern South America and South Africa. Subspecies unknown.
Status One record, in 1953.

Black-headed Gull
Chroicocephalus ridibundus   41cm
Local name Galleguito de Cabeza Negra.
Taxonomy Monotypic.
Description Non-breeding adult has pale grey mantle, dark ear-spot, white tail, red bill and legs. In flight shows distinct white wedge on leading edge of outer wing; underside of outer wing is dark smoke-grey. Breeding plumage has brown hood and white eye-ring. First-winter birds have pale dark-tipped bill and pale legs; dark brown carpal bar and tail-band.
Range Breeds throughout northern Europe and Asia, wintering in the Americas along Atlantic coast of North America from Labrador to New York, sparingly as far as northern South America.
Status Four records between 1964 and 2000.

Sabine's Gull
Xema sabini 34cm
Local name Galleguito de Cola Ahorquillada.
Taxonomy Monotypic.
Description Non-breeding adult has greyish half-hood on back of the head; upperwing divided into boldly contrasting triangles of white, grey and black. Bill black with yellow tip; legs black; tail deeply forked. Breeding bird has grey hood. Juvenile has entirely black bill; back and most of inner wing dark brown, wingtips black and remainder of wing white, giving a tricoloured appearance; broad black band on tail tip.
Range Breeds on Spitsbergen east to Taimyr Peninsula, and north-eastern Siberia to Alaska, Arctic Canada to Greenland, and winters in south-east Atlantic off south-west Africa and in eastern Pacific off north-west South America.
Status Three records offshore between 1954 and 1999.

Black-legged Kittiwake
Rissa tridactyla 43cm
Local name Gallego Patinegro.
Taxonomy Polytypic (2).
Description Non-breeding adult is white with grey wash on nape, distinct black ear-spot, light grey mantle, black wingtips with no white spots. Tail slightly forked. Bill yellow; legs black. Breeding plumage resembles winter adult plumage, with head completely white. First-winter has dark bars on lower hindneck and wings, forming M pattern, and dark band on tail tip; black bill. Flight fast, with rapid wingbeats.
Range Breeds in northern North America, east through northern Europe to Taimyr Peninsula; winters in northern Atlantic Ocean.
Status Seven records from shores of Cuba between 1946 and 2009.

Arctic Tern
Sterna paradisaea 39cm
Local name Gaviota Ártica.
Taxonomy Monotypic.
Description Breeding adult has black cap; bill and very short legs entirely red, grey underparts; upperwing appears almost uniformly grey, outer wing translucent with narrow black trailing edge on underside; tail deeply forked with dark edges. Non-breeding adult and immature have white forehead and black bill.

Range Breeds in Iceland, the British Isles east across Scandinavia almost exclusively north of the Arctic Circle to the Bering Strait, Alaska to Greenland, and south to New York. Winters in Antarctica.
Status Two records, in 1950 and 1969.

Large-billed Tern
Phaetusa simplex 38cm

Local name Gaviota de Pico Amarillo.
Taxonomy Monotypic.
Description Heavy, long yellow bill; cap black, dark grey back, white underparts, tail short and slightly forked, legs yellowish-grey. Upperwing has bold tricoloured pattern of black primaries, white secondaries, and grey inner wing. Juvenile resembles adult but with brownish spots, duller bill and streaks on head.
Range Breeds in South America.
Status Three records between 1909 and 2019.

Dovekie (Little Auk)
Alle alle 20cm

Local name Pingüinito.
Taxonomy Polytypic (2).
Description Small and stocky with large head. Non-breeding adult is black above, white below, breast may show brownish wash. Bill and tail very short. Breeding adult has head, throat and breast black; blackish scapulars with white streaks.
Range Breeds in Greenland and Iceland to Jan Mayen, Bear Island and Novaya Zemlya; winters at sea in North Atlantic and Bering Sea, exceptionally southern Florida, Bermuda and Grand Bahama. Breeds on Arctic coasts, wintering in North Atlantic.
Status At least 16 records between 1929 and 1985.

Long-eared Owl
Asio otus 38cm

Local name Búho.
Taxonomy Polytypic (4).
Description Dark brown above, spotted with darker brown; cream-coloured breast and abdomen, streaked dark brown. Yellow eyes in tawny-brown facial disc outlined in black; long ear-tufts. Long wings may extend beyond tail. In flight, shows buff underwing with black wrist mark.
Range *A. o. wilsonianus* breeds in Canada south to southern United States.
Status One record, in 1932.

Eastern Whip-poor-will
Antrostomus vociferus   24cm
Local name Guabairo Chico.
Taxonomy Monotypic.
Description Mottled dark greyish-brown, with broad, dark median crown-stripe on large head, throat black with contrasting white or buff U-shaped necklace. Short tail, three outer tail feathers tipped extensively white in male, buffy and more restricted in female. Rounded wings.
Range Breeds in southern Canada across eastern United States and winters in Gulf coast states and south through eastern Mexico to Honduras.
Status Four records, between 1932 and 2014.

Vermilion Flycatcher
Pyrocephalus rubinus   15cm
Local name Bobito Bermellón.
Taxonomy Polytypic (12).
Description Male has bright red underparts, and red on head contrasts with black mask from eye to nape, wings, back and tail. Short black bill and short tail; faint wingbars. Juvenile male has dense streaked breast and reddish on abdomen. Female has brown upperparts, pale supercilium and forehead, dark ear-patch; streaked below with pinkish wash on lower belly.
Range Breeds from south-western United States to northern Argentina, with southernmost populations moving north in austral winter to northern South America.
Status One record, in 2016. Unknown subspecies.

Alder Flycatcher
Empidonax alnorum   15cm
Local name Bobito de Alder.
Taxonomy Monotypic.
Description Virtually identical to Willow Flycatcher; only safely differentiated by voice. Brownish-olive above, eye-ring indistinct or lacking, wingbars whitish and conspicuous, whitish throat and pale olive breast, pale yellow belly, legs black, and mandible yellow. Juvenile has buff wingbars and indistinct eye-ring.
Range Breeds in Alaska, Canada, and north-eastern United States, wintering in South America.
Status Two records, in 1966 and 1967.

Tropical Kingbird
Tyrannus melancholicus 23cm
Local name Pitirre Tropical.
Taxonomy Polytypic (3).
Description Greenish back, buffy edges to wing-coverts, pale grey head with blackish through eye and ear-coverts, white throat, pale grey breast, yellow abdomen and undertail-coverts. Bill large and black, tail brownish-black and notched. Juvenile is duller, with paler head and whiter throat; crown patch is reduced or absent.
Range Breeds in south-western United States, Middle America to southern South America; winters in breeding range. *T. m. satrapa* breeds in southern United States to northern Colombia and Venezuela; Leeward Antilles, Trinidad, Tobago and Grenada.
Status Five records, between 19th century and 2017.

Cassin's Kingbird
Tyrannus vociferans 23cm
Local name Pitirre de Cassin.
Taxonomy Polytypic (2).
Description Adult has dark grey head and breast, contrasting with white throat, and yellow abdomen and undertail-coverts. Short, dark brown tail, notched or squared-tipped, pale on tip and on outermost rectrices. Juvenile is duller with browner upperparts.
Range Western and central United States to central Mexico, wintering south to western Guatemala.
Status One record, in 2017.

Fork-tailed Flycatcher
Tyrannus savana 33–41cm
Local name Bobito de Cola Ahorquillada.
Taxonomy Polytypic (4).
Description Black head, flight feathers and tail; grey mantle and wing-coverts, white underparts. Adult male's forked tail has very long dark tail-streamers with narrow white edges; female's tail shorter. Juvenile has brownish wash on head and mantle, and shorter tail.
Range Breeds in Middle and South America; nomadic and partly migratory; many winter in South America and irregularly to the southern Lesser Antilles.
Status. Two records, only one dated (1952). Unknown subspecies.

House Crow
Corvus splendens   42cm
Local name Cuervo.
Taxonomy Polytypic (5).
Description Blackish-slate plumage with glossy black head and rump to tail; greyish from nape to mantle, neck and sides of breast; prominent bill and legs black.
Range Native from south-eastern Iran to extreme southern China, Myanmar and Thailand, but has also spread through ship-assisted passage to Indian Ocean and Red Sea regions.
Status One record, an individual stayed in the same area for five years, between 2007 and 2011. Subspecies unknown.

Caribbean Martin
Progne dominicensis   20cm
Local name Golondrina del Caribe.
Taxonomy Monotypic.
Description Adult has entirely dark metallic purplish-blue upperparts, throat, upper breast and sides; blackish wings and forked tail; white lower breast to undertail-coverts. Female has paler upperparts, greyish-white throat, and greyish-brown upper breast and sides.
Range Greater Antilles (except Cuba), Lesser Antilles and Tobago; probably winters mainly in northern South America, including eastern Brazil.
Status One record, in 2011.

House Wren
Troglodytes aedon   12cm
Local name Troglodita Americano.
Taxonomy Polytypic (32).
Description Brown above, paler below; wings, sides and tail finely barred dark brown. Faint pale supercilium; bill thin and slightly downcurved. Juvenile has bright rufous rump and less distinct barring.
Range *T. a. aedon* breeds in North America, wintering south of breeding range to Florida, the Gulf coast and Mexico.
Status One record, in 1964.

Northern Wheatear
Oenanthe oenanthe 15cm
Local name Tordo Ártico.
Taxonomy Polytypic (4).
Description Breeding male is grey above and cinnamon-buff below; white supercilium; black mask, wings and broad terminal band on tail. Female similar, with brownish upperparts; brownish-grey mask and wings. Both sexes have large white rump patch extending to base of tail. Male in autumn resembles female. Juvenile is cinnamon-buff below with buffy supercilium.
Range Breeds in North American Arctic, northern Europe, Asia and northern Africa. Winters in Africa and Asia. Casual along United States eastern seaboard on migration.
Status One record, in 1903. Subspecies unknown.

Eastern Bluebird
Sialia sialis 18cm
Local name Azulejo Pechirojo.
Taxonomy Polytypic (8).
Description Male is blue above, rusty-red below, with white lower belly and undertail-coverts. Female similar but duller. Juvenile has brown upperparts and mottled underparts. Wings and tail have a trace of blue.
Range Breeds from southern Canada and the eastern United States south through Middle America to northern Nicaragua and Bermuda, with northernmost populations wintering south to north-eastern Mexico.
Status In 19th century was considered a rare winter visitor on mainland Cuba. Eleven records, between 1920 and 2012.

Hermit Thrush
Catharus guttatus 18cm
Local name Tordo de Cola Carmelita.
Taxonomy Polytypic (9).
Description Brownish-olive upperparts with contrasting rufous tail and reddish-brown primaries. Whitish eye-ring and white throat, upper breast densely smudged with dark brown spots; lower breast and abdomen white. Juveniles often have buff-tipped greater secondary coverts, forming single wingbar. Flicks wings and tail.
Range Breeds throughout northern and western United States and southern Canada north to Alaska, wintering from western and southern United States to Guatemala and El Salvador.
Status Two records, in 1995 and 2001. Subspecies unknown.

Brown Thrasher
Toxostoma rufum 29cm
Local name Sinsonte Colorado.
Taxonomy Polytypic (2).
Description Bright rufous above with long rufous tail; creamy-white below heavily streaked dark brown. White wingbars. Adult eye yellow; grey in juvenile.
Range *T. r. rufum* breeds in southern Canada and central and eastern United States, wintering within the United States.
Status Three records between 1963 and 2001.

Starling
Sturnus vulgaris 15cm
Local name Estornino.
Taxonomy Polytypic (13).
Description Stocky with short tail. Adult non-breeding is glossy black heavily spotted with white, bill black and sharply pointed. Adult breeding is glossy black with purple iridescence on head, neck and coverts, white spots on mantle and abdomen; bright yellow bill and pink legs. Juvenile is brownish-grey with dark bill.
Range Old World species. *S. v. vulgaris* is introduced species breeding in North America and the Caribbean.
Status Five records between 1955 and 1987.

American Pipit
Anthus rubescens 17cm
Local name None.
Taxonomy Polytypic (3).
Description Head small, short pointed bill, white to buffy supercilium, broad malar stripe, distinct wingbars. Tail long with white outer tail feathers; bobs tail while feeding on ground. Legs generally dark. Breeding plumage may show slight streaks on back; breast and flanks buffy with streaking variable or absent; underparts densely streaked in winter.
Range Breeds in North America, from Alaska to Maine, and in Asia, west to the Taimyr Peninsula and south to Transbaikalia; winters throughout the southern United States and Middle America.
Status Three records between 2006 and 2017.

Orange-crowned Warbler
Leiothlypis celata 13cm
Local name Bijirita de Coronilla Anaranjada.
Taxonomy Polytypic (4).
Description Olive above with faint but distinct greyish wash on head, paler below lightly and diffusely streaked; short supercilium and faint dark eye-line and indistinct yellowish eye-ring; wings unmarked or with very indistinct wingbars; faint blurred streaks on side of breast and sides; undertail-coverts yellow, outer retrices edged whitish on inner web. A concealed orange patch on crown, very seldom visible. Female similar to male, but smaller crown patch (noticeable only in the hand). First-winter has greyish head, olive-grey upperparts, and narrow yellowish wingbars.
Range Breeds in northern and western North America, wintering from southern United States to Guatemala.
Status Twelve dated records between 19th century and 2017.

Black-throated Gray Warbler
Setophaga nigrescens 13cm
Local name Bijirita Gris de Garganta Negra.
Taxonomy Monotypic.
Description Black head boldly marked with white stripe and yellow loral spot. Grey above with white wingbars; white underparts. Male has black throat and breast, and bold black streaks on sides. Female similar but duller above; white chin; throat white with blackish-band or may be white mixed with blackish. Immature male like female. Immature female has little black on throat and breast and is brownish-grey above.
Range Breeds in south-western Canada and western United States, wintering from California and south-western United States to Mexico.
Status One record, in 1997. First record for Caribbean.

Townsend's Warbler
Setophaga townsendi 13cm
Local name Bijirita de Townsend.
Taxonomy Monotypic.
Description Male has blackish crown and ear-patch bordered in yellow; olive upperparts streaked blackish. Black chin and throat have bold black streaks at sides. Yellow on underparts is more extensive to mid-belly. Female is duller with yellow throat and upper breast mottled blackish, finer streaks on upperparts and sides.

Range Breeds in Alaska south to Oregon and Montana; winters from extreme south-western British Columbia to western and central Mexico south to Costa Rica.
Status One record, in 2015.

Pine Warbler
Setophaga pinus 14cm
Local name Bijirita de Pinos.
Taxonomy Polytypic (4).
Description Long thick bill; tail seems long due to short undertail-coverts; black feet and legs. Male has unstreaked olive upperparts and conspicuous white wingbars; throat and breast yellow, sides faintly streaked olive; abdomen and undertail-coverts white. Female similar but duller, with indistinct streaks on sides. Juvenile mostly brownish, washed green on back, with yellowish or buffy underparts.
Range Breeds in eastern North America, wintering to southern United States. Breeding resident in Bahamas and Hispaniola.
Status Eight records between 1964 and 2001.

Kirtland's Warbler
Setophaga kirtlandii 15cm
Local name Bijirita de Kirtlandi.
Taxonomy Monotypic.
Description Male has blue-grey upperparts with black streaks on back, faint wingbars and long tail; black lores and broken white eye-ring; underparts yellow with black streaks on sides and white undertail-coverts. Female similar with brownish-grey upperparts and grey lores.
Range Breeds in northern United States; winters in Bahamas and on Turks and Caicos Islands.
Status Two records, in 2004 and 2017.

Connecticut Warbler
Oporornis agilis 14cm
Local name Bijirita de Connecticut.
Taxonomy Monotypic.
Description Distinct large eye with white or buff eye-ring (can be broken at rear), brownish upperparts, yellow abdomen and undertail-coverts, no wingbars, long bill (pale mandible), pale legs and short tail. Male has grey head and upper breast forming a hood, becoming darker on upper breast, paler on throat. Female is duller overall with brownish or greyish-brown hood. A ground-

dweller, walks rather than hops.
Range Breeds in northern North America; winters in north-central South America.
Status Two records, in 1968 and 2001.

Mourning Warbler
Geothlypis philadelphia 13cm
Local name Bijirita de Cabeza Gris.
Taxonomy Monotypic.
Description Male has plain olive upperparts, with dark grey hood to upper breast, broadly mottled black on breast; black lores, incomplete eye-ring; black maxilla and pink mandible; underparts otherwise completely bright yellow. Female is similar, with pale grey or brownish-grey hood, pale throat, no black on breast, and almost complete eye-ring. First-winter male resembles adult female with yellower throat and diffuse hooded effect; first-winter female is duller. All plumages have pale pink feet that are characteristic of the genus. Walks rather than hops.
Range Breeds in northern North America; winters from southern Nicaragua to northern South America.
Status Two records, in 2016 and 2018.

Saffron Finch
Sicalis flaveola 14cm
Local name Gorrión Azafrán.
Taxonomy Polytypic (5).
Description Adult entirely bright yellow with orange crown, conical bill. Female paler with little orange on forecrown. Juvenile duller with greyish head, streaked upperparts, white underparts with yellow breast-band.
Range South America; in West Indies, introduced to Jamaica and Puerto Rico.
Status Two records, in the 19th century and 1996.

Green-tailed Towhee
Pipilo chlorurus 18cm
Local name Gorrión de Cola Verde.
Taxonomy Monotypic.
Description Dark olive upperparts; grey head with reddish-brown crown, white chin and black malar stripe, white loral spot. Grey below with rather sharply defined white throat; bill short and conical.
Range Breeds in western United States, wintering to Mexico.
Status One record, in 1964. Another undated from Cayo Coco.

Rufous-collared Sparrow
Zonotrichia capensis 15.0–16.5cm
Local name None.
Taxonomy Polytypic (28).
Description Grey head with black stripes on crown, bright rufous band across neck (often extended to upper breast sides), back brown and streaked, all feathers broadly fringed brown. White throat and upper broad black breast band; rest of underparts greyish white, with pale buffy sides. Bill greyish. Juvenile duller, head striped brown and greyish, lacks black breast band and rufous on neck, underparts streaked.
Range Middle and South America, and Hispaniola.
Status One record, in 1935. Either originated from captivity or was deliberately introduced.

Lapland Longspur (Lapland Bunting)
Calcarius lapponicus 16cm
Local name None.
Taxonomy Polytypic (5).
Description Non-breeding bird has black on side of face (ear-covert area), which form broad black triangle outlining buffy ear-coverts patch, and small conical orange-yellow bill, short dark tail with white outer feathers. Male has bright rufous-brown back streaked blackish, breast heavily mottled with black. Female shows less rufous and less distinct streaks on buffy breast. Both have distinct rufous-brown patch between faint wingbars; white underparts and heavily streaked sides. Breeding male has chestnut nape, and black crown, face and breast outlined with broad white stripe. Female lacks the extensive black and contrasting chestnut nape-patch.
Range *C. l. lapponicus* breeds in North America from Alaska east to Ontario, and widely at Arctic latitudes across the Holarctic. Winters further south, including in southern United States.
Status Two records, in 2016 and 2019.

Black-headed Grosbeak
Pheucticus melanocephalus   21cm
Local name Degollado de Cabeza Negra.
Taxonomy Polytypic (2).
Description Male has black head with massive conical bill (maxilla darker, mandible paler); orange overall; wings blackish with broad white patches. Female has dark brown head, white or buffy long supercilium, upperparts brown with streaks, white wingbars; underparts buffy with fine streaks on sides and flanks. First-winter male is orange-buff below with striped head. In flight both sexes show yellow wing-linings.
Range South-western Canada to south-western United States and western Mexico.
Status One record, in 2007. Subspecies unknown.

Lazuli Bunting
Passerina amoena   14cm
Local name Mariposa Azul.
Taxonomy Monotypic.
Description Male breeding has turquoise-blue head, back and rump, dark wings with broad white wingbars and wider upper-bar; breast and sides orange-chestnut, rest of underparts white; bill small and conical, and dark lores. Non-breeding male is duller, with brown edges to upperparts and throat. Female brown above, bluish on rump and tail, pale buffy wingbars, greyish throat and unstreaked buffy breast. Juvenile similar to female, with faint streaks on breast and brown rump.
Range Breeds in central and western North America, wintering to Mexico.
Status Five records between 1960 and 2013.

Yellow-headed Blackbird
Xanthocephalus xanthocephalus   25cm
Local name Mayito de Cabeza Amarilla.
Taxonomy Monotypic.
Description Male black with orange-yellow head and breast, black lores and white wing-covert patch. Female uniform dusky-brown; dull yellow face, throat and breast; brownish lores, lower breast streaked and wing-patch absent. Juvenile male resembles female, with darker head and lores and more extensive yellow on head.
Range Breeds in western North America, wintering to Mexico.
Status Six records between 19th century and 2015.

Brown-headed Cowbird
Molothrus ater   19cm
Local name Totí Americano.
Taxonomy Polytypic (3).
Description Male black with green-and-purple metallic sheen and dark brown head. Female greyish-brown, somewhat darker above and faintly streaked below. Both sexes have short, square tail and deep conical bill. Juvenile is similar to female, conspicuously streaked below. Moulting juvenile male has an irregular pattern of black blotches on grey.
Range Breeds in North America to Mexico, wintering in southern parts of the breeding range.
Status Two records, both in 1960. Subspecies *M. a. ater*.

Hooded Oriole
Icterus cucullatus   20cm
Local names Turpial de Capucha.
Taxonomy Polytypic (5).
Description Breeding adult male is yellowish-orange, black lores to upper breast, upper back, wings and tail. Non-breeding male has buffy tips on back. Female is entirely brownish-grey to greenish-yellow with orange tint; paler below. Both sexes have long, slightly decurved, pointed bill, wingbars (male with broad white median coverts) and long, markedly graduated tail. Juvenile similar to female; immature male has black throat, and pale lores.
Range Breeds in south-western United States and Mexico, wintering within the breeding range.
Status Four records between 19th century and 2009. Subspecies unknown.

Appendix B
Introduced species breeding on Cuba

Muscovy Duck
Cairina moschata Male 80cm • Female 63cm
Local name Pato Doméstico.
Taxonomy Monotypic.
Description A black duck with large white wing-patches and underwing-coverts. Male has a bare, knobby red face. Female duller; may lack facial knobs. Juvenile black with small white spot on upperwing. Domestic varieties can be white, black or mottled. Escaped birds often become semi-wild.
Voice Usually silent, but may emit hissing noises or low quacks.
Similar species None.
Habitat and behaviour Ponds, slow-moving rivers. Breeds May–July. Lays 8–9 greenish-white eggs in tree hollows. Feeds on aquatic vegetation, fruits, insects and small fish.
Range Northern Mexico south through Central America to Argentina.
Status Widely introduced.

Ring-necked Pheasant
Phasianus colchicus Male 84cm • Female 53cm
Local name Faisán de Collar.
Taxonomy Polytypic (30).
Description Male large, multicoloured and iridescent. Green to purple head, with conspicuous bare red skin on face and usually a white neck-ring. Long and pointed tail. Female smaller, mottled brown, with shorter tail. Both sexes have short rounded wings and when flushed arise with a sudden burst of loud whirring.
Voice Male gives loud crowing *haa-haak*, followed by a whirr of wings. Both sexes emit croaking alarm notes.
Similar species None.
Habitat and behaviour Grassy open country. Breeding undescribed on Cuba. Omnivorous.
Range Asia. Widely introduced in Europe, Hawaiian islands, New Zealand and North America.
Status Introduced with very limited success at Guanahacabibes and Topes de Collantes; common at Los Indios, Isle of Pines.

Helmeted Guineafowl
Numida meleagris   56cm

Local name Guineo.
Taxonomy Polytypic (9).
Description Greyish-black, with plumage entirely and very finely spotted with white. Wholly or partially white birds are occasionally encountered, usually around farms with other domestic birds. Head and neck naked. An excellent runner that only rarely takes flight.
Voice A loud, grating *cherrrr* or *kek-kek-kek-krrrr* alarm call; also, a loud *Pas-cual, Pas-cual*.
Similar species None.
Habitat and behaviour Common in pastures, savannas, grassy fields with low bushes. Breeds on ground. Lays up to 16 pale brown eggs. Feeds on insects, molluscs and shoots.
Range Africa.
Status Introduced during the slave trade and established throughout mainland Cuba near farms and towns.

Rock Pigeon
Columba livia   33–36cm

Local name Paloma Doméstica.
Taxonomy Polytypic (13).
Description Individuals show wide colour variation (brown, white, multicoloured and dark); the most common form has a dark bluish-grey head and nape, iridescent greenish-purple sheen on sides of neck, pale grey back and two broad black wingbars, orange iris, white cere, black bill and reddish-plum legs. In flight shows white rump and black terminal band on tail.
Voice Call a low series of cooing notes.
Similar species None.
Habitat and behaviour Forages for seeds on the ground in urban areas and rubbish tips. Breeds throughout the year in rough stick nest on ledges of low buildings; produces a clutch of two white eggs.
Range Introduced and established worldwide, including the West Indies.
Status Introduced, common.

House Sparrow
Passer domesticus   15cm
Local name Gorrión Doméstico.
Taxonomy Polytypic (12).
Description Non-breeding male has greyish crown, rump and tail, greyish-white cheek patch, black eye-line and black throat and breast edged grey; rest of underparts greyish; upperparts rufous and brown with black streaks; pale legs and heavy short, dark yellowish bill. Breeding male has bright rufous upperparts and nape, black breast, black bill, broad white band on wing. Female has brownish crown with buff supercilium and lores, bill horn, upperparts brown streaked black, and buffy-grey underparts; juvenile similar to female but duller.
Voice *Chirrup*, repeated.
Similar species None.
Habitat and behaviour Urban areas, second growth. Breeding January–August. Nest is rather large and messy, vaguely cup-shaped. Lays 3–4 white eggs, very variably marked with spots, speckling or small blotches of grey, greenish-grey, black or brown. Omnivorous.
Range Native to Eurasia and North Africa. Introduced worldwide. *P. d. domesticus* introduced and breeding in many cities and towns in the Americas, and colonised northern Bahamas, Greater Antilles and Cayman Islands. Cosmopolitan.
Status Very common breeding resident.

Scaly-breasted Munia
Lonchura punctulata   11.5cm
Local name Damero.
Taxonomy Polytypic (9).
Description Cinnamon head and upperparts; scalloped underparts; blackish bill. Juvenile unmarked below.
Voice A soft whistle *peet*.
Similar species Juvenile Tricoloured Munia is duller below with a paler bill.
Habitat and behaviour Second growth, grasslands and cultivated areas. Breeding season reported to be from April–August, but no other details on Cuba. Forage in flocks.
Range India, Nepal, southern China and through south-west Asia to Indonesia and Philippines. In Americas introduced in the Greater Antilles, Guadeloupe and south-western USA (California).
Status Already common. It is still expanding rapidly on the mainland.

Tricoloured (Chestnut) Munia
Lonchura malacca   11.5cm
Local name Monja Tricolor.
Taxonomy Polytypic (9).
Description Strikingly marked with black hood, bluish bill, cinnamon back and black patch on white abdomen. Juvenile brown.
Voice A nasal *honk*.
Similar species Juvenile Scaly-breasted Munia has browner underparts and a darker bill.
Habitat and behaviour Wetlands, grasslands. Breeding April–August. Builds bulky nest with entrance hole in the side. Lays 4–5 white eggs. Known also to nest in sugar cane fields in Puerto Rico.
Range India through South-East Asia to Philippines. Introduced in Puerto Rico, reaching Cuba by unknown means, possibly hurricanes.
Status Widespread and locally common.

Chestnut Munia
Lonchura atricapilla   11–12cm
Local name Monja Castaña.
Taxonomy Polytypic (8).
Description Chestnut-brown with glossy black hood; bluish-grey, thick and conical bill; and rump and tail deep cinnamon-orange. Central belly and tail-coverts black. Female is duller, with less glossy hood and smaller black patch on belly. Juvenile has bluish-grey bill with dark edging to mandible, warm brown upperparts and buff underparts.
Voice Calls a thin, nasal honk, also *peet* or *pink, pink*.
Similar species Adults unmistakable. Juveniles very similar to other juvenile mannakins.
Habitat and behaviour Second growth, marshes and coasts, tall grass bordering dense vegetation, grassland, and urban areas. Typically seen in flocks, foraging on the ground or while perched.
Range North India, east to south China and South-East Asia. Introduced in the West Indies (Jamaica, Martinique), southern Europe, Japan, Hawaiian Islands and Ecuador (Guayas).
Status Cuba. Apparently an established introduction, recorded from Holguin province in 2013.

Appendix C
Species likely to be extinct on the Cuban archipelago

Cuban Kite has heavy hooked bill. Female (left) has conspicuous reddish bars on whitish underparts; male (right) is ash-coloured, barred below, mostly with grey.

Adult Zapata Rail has olive bill with red base, red legs and grey underparts.

Male Ivory-billed Woodpecker has red on crest; female has all-black crest. Young male can be seen in the cavity.

Appendix D
Endemic species

Common Name	Scientific Name	Archipelago distribution		
		Cuba	Isle of Pines	Cays
Cuban Kite	*Chondrohierax wilsonii*	X		
Gundlach's Hawk	*Accipiter gundlachi*	X		X
Cuban Black Hawk	*Buteogallus gundlachii*	X	X	X
Zapata Rail	*Cyanolimnas cerverai*	X		
Gray-fronted Quail-Dove	*Geotrygon caniceps*	X		
Blue-headed Quail-Dove	*Starnoenas cyanocephala*	X	X[a]	
Cuban Parakeet	*Psittacara euops*	X		
Bare-legged Owl	*Margarobyas lawrencii*	X	X	X
Cuban Pygmy-Owl	*Glaucidium siju*	X	X	X
Cuban Nightjar	*Antrostomus cubanensis*	X	X	X
Bee Hummingbird	*Mellisuga helenae*	X	X	
Cuban Trogon	*Priotelus temnurus*	X	X	X
Cuban Tody	*Todus multicolor*	X	X	X
Cuban Green Woodpecker	*Xiphidiopicus percussus*	X	X	X
Fernandina's Flicker	*Colaptes fernandinae*	X		
Giant Kingbird	*Tyrannus cubensis*	X	X	
Cuban Vireo	*Vireo gundlachii*	X	X	X
Cuban Palm Crow	*Corvus minutus*	X		
Zapata Wren	*Ferminia cerverai*	X		
Cuban Gnatcatcher	*Polioptila lembeyei*	X		X
Cuban Solitaire	*Myadestes elisabeth*	X	X[a]	
Yellow-headed Warbler	*Teretistris fernandinae*	X	X	X
Oriente Warbler	*Teretistris fornsi*	X		X
Cuban Bullfinch	*Melopyrrha nigra*	X	X	X
Cuban Grassquit	*Tiaris canorus*	X		
Zapata Sparrow	*Torreornis inexpectata*	X		X
Red-shouldered Blackbird	*Agelaius assimilis*	X	X	
Cuban Blackbird	*Ptiloxena atroviolacea*	X		X[b]
Cuban Oriole	*Icterus melanopsis*	X	X	X

X[a]: Possibly extirpated
X[b]: Not documented

Cuban Green Woodpecker.

Appendix E
Endemic subspecies

Common name	Scientific name	Archipelago distribution		
		Cuba	Isle of Pines	Cays
Sharp-shinned Hawk	*Accipiter striatus fringilloides*	X		
Gundlach's Hawk*	*Accipiter gundlachi gundlachi*	X	X	X
	Accipiter g. wileyi	X		
Broad-winged Hawk	*Buteo platypterus cubanensis*	X		
American Kestrel	*Falco sparverius sparverioides*	X	X	X[a]
Northern Bobwhite	*Colinus virginianus cubanensis*			X[a]
King Rail	*Rallus elegans ramsdeni*	X	X	
Sandhill Crane	*Grus canadensis nesiotes*	X	X	
Cuban Parrot	*Amazona leucocephala leucocephala*	X	X	
Great Lizard-Cuckoo	*Coccyzus merlini merlini*	X		
	Coccyzus m. decolor		X	
	Coccyzus m. santamariae			X[b]
Cuban Pygmy-Owl*	*Glaucidium siju siju*	X		
	Glaucidium s. vittatum	X	X	
Stygian Owl	*Asio stygius siguapa*	X	X	X
Cuban Nightjar*	*Antrostomus cubanensis cubanensis*	X		X
	Antrostomus c. insulaepinorum		X	
Antillean Palm-Swift	*Tachornis phoenicobia iradii*	X	X	
Cuban Trogon*	*Priotelus temnurus temnurus*	X		X
	Priotelus t. vescus		X	
West Indian Woodpecker	*Melanerpes superciliaris superciliaris*	X		X
	Melanerpes s. murceus		X	X
Cuban Green Woodpecker*	*Xiphidiopicus percussus percussus*	X		X
	Xiphidiopicus p. insulaepinorum		X	X
Northern Flicker	*Colaptes auratus chrysocaulosus*	X		X
Ivory-billed Woodpecker	*Campephilus principalis bairdii*	X		
Cuban Pewee	*Contopus caribaeus caribaeus*	X	X	X
	Contopus c. morenoi	X		X[c]
	Contopus c. nerlyi			X[d]

		a	b	c	d
Cuban Vireo*	Vireo gundlachii gundlachii	X	X		X
	Vireo g. orientalis	X			X
Loggerhead Kingbird	Tyrannus caudifasciatus caudifasciatus	X	X		X
Thick-billed Vireo	Vireo crassirostris cubensis				X
Cuban Solitaire*	Myadestes elisabeth elisabeth	X			
	Myadestes e. retrusus				X[e]
Red-legged Thrush	Turdus plumbeus rubripes	X	X		X
	Turdus p. schistaceus	X[e]			
Western Spindalis	Spindalis zena pretrei	X	X		X
Zapata Sparrow*	Torreornis inexpectata inexpectata	X[f]			
	Torreornis i. sigmani	X[g]			
	Torreornis i. varonai				X
Tawny-shouldered Blackbird	Agelaius humeralis humeralis	X	X		X
	Agelaius h. scopulus				x[h]
Eastern Meadowlark	Sturnella magna hippocrepis	x	x		x
Greater Antillean Grackle	Quiscalus niger gundlachii	x			x
	Quiscalus n. caribaeus	x	x		x

a: Cayo Sabinal
b: Cayo Santa María, Cayo Coco, Cayo Romano
c: Cays south of Villa Clara
d: Jardines de la Reina Archipelago
e: Eastern Cuba
f: Ciénaga de Zapata
g: South-east Guantánamo
h: Cayo Cantiles
* Endemic species

BIBLIOGRAPHY

Acosta Cruz, M. & Mugica Valdés, L. 2006. *Aves Acuáticas en Cuba*. Facultad de Biología, Universidad de La Habana, Havana, Cuba.

American Ornithologists' Union (AOU). 1983. *Check-list of North American Birds*, 6th edition. American Ornithologists' Union, Washington, D.C.

American Ornithologists' Union (AOU). 1998. *Check-list of North American Birds*, 7th edition. American Ornithologists' Union, Lawrence, Kansas.

Arredondo, O. 1984. Sinópsis de las Aves Halladas en Depósitos Fosilíferos Pleisto-holocénico de Cuba. *Rep. Invest. Inst. Zool.* 17: 1–35.

Bond, J. 1956. *Check-list of the Birds of the West Indies*. Acad. Nat. Sci., Philadelphia.

Borhidi, A. 1996. *Phytogeography and Vegetation Ecology of Cuba*, 2nd edition. Akadémiai Kiadó, Budapest.

Bradley, P. E & Rey-Millet, Y-J. 2013. *Birds of the Cayman Islands*. Christopher Helm, London.

Chesser, R. T., Burns, K. J., Cicero, C., Dunn, J. L., Kratter, A. W., Lovette, I. J., Rasmussen, P. C., Remsen, J. V., Rising, J. D., Stotz, D. F. & Winker, K. 2016. Fifty-seventh Supplement to the American Ornithologists' Union *Check-list of North American Birds*. *Auk* 133: 544–560.

Dickinson, E. C. & Christidis, L. (eds.) 2014. *The Howard and Moore Complete Checklist of the Birds of the World*. Vol. 2, 4th edition. Aves Press, Eastbourne.

Dickinson, E. C. & Remsen, J. V. (eds.) 2013. *The Howard and Moore Complete Checklist of the Birds of the World*. Vol. 1, 4th edition. Aves Press, Eastbourne.

Garrido, O. H. & García Montaña, F. 1975. *Catálogo de las Aves de Cuba*. Acad. Cienc. Cuba, La Habana.

Garrido, O. H. & Kirkconnell, A. 2000. *Field Guide to the Birds of Cuba*. Cornell Univ. Press, Ithaca, NY.

Garrido, O. H., Wiley, J. W., Kirkconnell, A., Bradley, P. E., Gunther-Calhoun, A. & Rodríguez, D. 2014. Revision of the Endemic West Indian Genus *Melopyrrha* from Cuba and the Cayman Islands. *Bulletin of the British Ornithologists' Club* 134: 120-128.

Gundlach, J. 1876. *Contribución a la Ornitología Cubana*. Imprenta La Antilla, La Habana.

Keith, A. R., Wiley, J. W., Latta, S. C. & Ottenwalder, J. A. 2003. *The Birds of Hispaniola: an Annotated Checklist*. BOU checklist No. 21. British Ornithologists' Union & British Ornithologists' Club, Tring.

Kirkconnell, A., Kirwan, G. M., Garrido, O. H., Mitchell, A. & Wiley, J. W. in press. *The Birds of Cuba: an Annotated Checklist*. British Ornithologists' Club, Tring.

Pregill, G. K., & Olson S. L. 1981. Zoogeography of West Indian Vertebrates in Relation to Pleistocene Climate Cycles. *Ann. Rev. Ecol. Syst.* 12: 75-98.

Ricklefs, R. E. & Bermingham, E. 2008. The West Indies as a Laboratory of biogeography and evolution. *Phil. Trans. R. Soc. B*. 363: 2393–2413.

Suárez, W. 2004. Biogeografía de las Aves Fósiles de Cuba. *En* Iturralde-Vinent, M. (ed), *Origen y Evolución del Caribe y sus biotas Marinas y Terrestres*. Centro Nacional de Información Geológica, La Habana, Cuba.

Wiley, J. W. 2000. A Bibliography of Ornithology in the West Indies. *Proc. West. Found. Vert. Zool.* 7: 1–817.

IMAGE CREDITS

All photographs taken by Yves-Jacques Rey-Millet with the exception of the following:

(t = top, b = bottom, l = left, r = right, c = centre, tl = top left, tr = top right, bl = bottom left, br = bottom right, cl = centre left, cr = centre right)

AGAMI Photo Agency/Alamy 227t, 354t; Alexandra Rudge/Getty 367t; All Canada Photos/Alamy 362b; Alvaro de Jesús 371tl, 371tr, b; Alvaro Jaramillo 98b; Arthur Morris/Getty 120b, 347c; Arturo Kirkconnell 14, 15, 16, 17bl, 17br, 18t, 18b, 19tr, 19cl, 19cr, 19bl, 19br, 20, 21t, 21bl, 21br, 22t, 22b, 101bl, 104bl, 110b, 118tl, 118tr, 118b, 125b, 176cl, 178bl, 179l, 181b, 185br, 195br, 195br, 196l, 196r, 197t, 202t, 202b, 213t, 214br, 232bl, 232br, 233tl, 233tr, 255t, 255b, 260t, 260bl, 261tr, 276bl, 317bl, 317br, 329bl, 335b; Arturo Kirkconnell Jr. front cover (br), 23, 26, 35, 50tl, 52b, 55tr, 55br, 56b, 69t, 86t, 90t, 93cr, 101t, 102t, 102br, 105br, 123l, 124b, 128br, 129br, 133tr, 136cr, 138l, 139b, 142b, 151b, 156t, 162r, 174b, 183t, 184bl, 189tl, 189tr, 189br, 193, 204t, 205t, 210b, 211t, 212bl, 212br, 237r, 239tl, 239tr, 244b, 245tl, 246t, 248tl, 248bl, 248br, 249tl, 250bl, 252b, 253tr, 254l, 254r, 261tl, 307tr, 309t, 309b, 311l, 312tr, 312b, 315bl, 330tl, 335tr, 335cr, 337t, 337b; Bertrando Campos/Getty 355t; Bill Clark 95t; Bill Draker/Getty 347t; Bill Gorum/Alamy 349c; Bill Raften/Getty 352b; blickwinkel/Alamy 4; Brendan Fogarty 250t; Brent Stephenson/Nature Picture Library 348b; Brian E. Small/VIREO 108; Bruce Hallet 74t, 77t, 82tl, 82cl, 82bl, 85t, 86b, 92b, 94t, 110t, 120t, 133br, 140t, 151t, 153tr, 153b, 160t, 168tl, 201b, 223, 241b, 298t, 316b, 340t; C. VIREO/G. McElroy 283t; Charles Melton/Alamy 319t; Claudio Contreras/Nature Picture Library 248c; David Sainsbury/Getty 354b; Denis Duflo/Getty 369b; Diana Robinson Photography/Getty 361c; Doug McSpadden/500px/Getty 356c, 365c; Ed Reschke/Getty 349l; Edwin 010/500px/Getty 351t; Elizabeth Riley 70br; Elizabeth W. Kearley/Getty 53bl; Erhard Nerger/Getty 346b; Frans Sellies/Getty 364b; G. Bartley/VIREO 251t; G. McEvoy/VIREO 262b; Gabriel Rojo/Nature Picture Library 346t; Gary Carter/Getty 359c; Gary Chalker/Getty 367b; Gary Fairhead/Getty 366t; Gerrit Vyn/Nature Picture Library 349b, 358b, 361b, 363t, 363b; Glen Tepke 44t, 44b, 82tr, 149t, 204b; Hanne & Jens Eriksen/Nature Picture Library 147; imageBROKER/Kevin Sawford/Getty 347b; imageBROKER/Michael Weber/Getty 369t; Its About Light/Design Pics/Getty 366b; James Mundy/Nature's Ark Photography/Alamy 256b; James Warwick/Getty 345c; jared Lloyd/Getty 350b; John Shaw/Nature Picture Library 352c; Jussi Murtosaari/Nature Picture Library 351c, 355c; Kevin Schafer/Getty 368t; Linda Kreuger/Getty 351b; Linda Kreuger/500px/Getty 360t; Loic Poidevin/Nature Picture Library 356t; Lynn M Stone/Nature Picture Library 348t; Manuel ROMARIS/Getty 357t; Mark L Stanley/Getty 354c; Mark Newman/Getty 345b; Markus Varesvuo/Nature Picture Library 344t; McPhoto/Lovell/Alamy 344b; Michael Nolan/Getty 343c; Michael S. Nolan/Alamy 172t; Michel & Gabrielle Therin-Weise/Alamy 363c; Mike Powles/Getty 346c, 353b, 355b, 356t, 358t; Minden Pictures/Alamy 343b, 362c; Moelyn Photos/Getty 350t; Nancy Norman 47t, 47b, 67b, 95cr, 95br, 101bl, 104cl, 121b, 126b, 150bl, 161t, 165t, 177b, 186t, 186b, 199t, 199b, 226t, 228l, 232t, 233b, 235t, 251b, 264t, 264b, 271t, 272b, 281br, 288bl, 290b, 298b, 304t, 305b, 306t, 319b, 320b, 324t; Neal Mishler/Getty 365b; Neil Bowman/FLPA 338t; Nigel Eve/Getty 360c; Nitat Termmee/Getty 370b; Patricia E. Bradley 52cl; R. Curtis/VIREO 206t; Remco Douma/Getty 350c; Richardom/Alamy 105bl; Rick & Nora Bowers/Alamy 356b; Robert L. Potts/Design Pics/Getty 365t; Robin Chittenden/Nature Picture Library 343t; Rolf Nussbaumer/Getty 360b, 362t; Ronai Rocha/Getty 357b; sandra stanbridge/Getty 359t, 364b, 368b; Susan Gary/Getty 53br; Thanit Weerawan/Getty 352t, 370t; Thomas Lazar/Nature Picture Library 342t; Tom Vezo/Nature Picture Library 357c; Tony Beck/500px/Getty 344c; Towsend Dickinson 272t, 274t, 290t, 361t; Vicki Jauron, Babylon and Beyond Photography/Getty 359b; Vince F/Alamy 342c; Westend61/Getty 353t.

INDEX OF ENGLISH NAMES

A
Anhinga 51
Ani, Smooth-billed 190
Auk, Little 355
Avocet, American 128

B
Bananaquit 306
Bittern,
　American 53
　Least 54
Blackbird,
　Cuban 333–4
　Red-shouldered 329–30
　Tawny-shouldered 331
　Yellow-headed 366
Bluebird, Eastern 359
Bobolink 328
Bobwhite, Northern 107
Booby,
　Brown 46
　Masked 344
　Red-footed 345
Bufflehead 349
Bullfinch,
　Cuban 311–12
　Indigo 325
　Lapland 364–5
　Lazuli 365
　Painted 326

C
Canvasback 82
Caracara, Crested 103
Catbird, Grey 267
Chat, Yellow-breasted 306
Chuck-will's-widow 201
Coot, American 116
Cormorant,
　Double-crested 50
　Neotropic 49
Cowbird,
　Brown-headed 366

　Shiny 336
Crake, Yellow-breasted 112
Crane, Sandhill 118
Crow,
　Cuban 246
　Cuban Palm 244–5
　House 358
Cuckoo,
　Black-billed 186
　Mangrove 188
　Yellow-billed 187
Curlew, Long-billed 350

D
Dickcissel 327
Diver, Great Northern 342
Dove,
　Eurasian Collared 172
　Mourning 175
　White-winged 173
　Zenaida 174
Dovekie 355
Dowitcher,
　Long-billed 147
　Short-billed 146
Duck,
　Masked 87
　Mottled 348
　Muscovy 367
　Ring-necked 83
　Ruddy 88
　Wood 73
Dunlin 144

E
Eagle, Bald 94
Egret,
　Cattle 61
　Great 56
　Reddish 60
　Snowy 57
Emerald, Cuban 207

F
Falcon, Peregrine 106

Finch, Saffron 363
Flamingo, American 69
Flicker,
　Fernandina's 221–2
　Northern 220
Flycatcher,
　Acadian 226
　Alder 357
　Fork-tailed 358
　Great Crested 228
　La Sagra's 229
　Scissor-tailed 235
　Vermilion 356
　Willow 227
　Yellow-bellied 226
Frigatebird,
　Magnificent 52

G
Gadwall 74
Gallinule, Purple 114
Gannet, Northern 345
Gnatcatcher,
　Blue-grey 257
　Cuban 258–9
Godwit,
　Hudsonian 250
　Marbled 350
Goose,
　Canada 347
　Greater White-fronted 346
　Snow 347
Greater Antillean 335
Grassquit,
　Black-faced 316
　Cuban 313–14
　Yellow-faced 315
Grebe,
　Least 42
　Pied-billed 42
Grosbeak,
　Black-headed 365
　Blue 324
　Rose-breasted 323
Ground-dove,
　Common 176

Guineafowl, Helmeted 368
Gull,
　Black-headed 354
　Bonaparte's 151
　Great Black-backed 155
　Herring 153
　Laughing 150
　Lesser Black-backed 154
　Ring-billed 152
　Sabine's 354

H
Harrier, Northern 95
Hawk,
　Broad-winged 100–1
　Cuban Black 99
　Gundlach's 96–7
　Red-tailed 102
　Sharp-shinned 98
　Swainson's 349
Heron,
　Great Blue 55
　Green 62
　Little Blue 58
　Tricolored 59
Honeycreeper, Red-legged 310
Hummingbird,
　Bee 209–10
　Ruby-throated 208

I
Ibis,
　Glossy 66
　Scarlet 345
　White-faced 346

J
Jacana, Northern 129
Jaeger,
　Long-tailed 353
　Parasitic 149
　Pomarine 149

K

Kestrel, American 104
Killdeer 125
Kingbird,
 Cassin's 357
 Eastern 230
 Giant 232–3
 Grey 231
 Loggerhead 234
 Tropical 357
Kingfisher, Belted 215
Kinglet, Ruby-crowned 256
Kite,
 Cuban 91, 371
 Mississippi 92
 Snail 93
 Swallow-tailed 92
Kittiwake, Black-legged 354–5
Knot, Red 137

L

Limpkin 117
Lizard-cuckoo, Great 189
Longspur, Lapland 364–5
Loon, Common 342

M

Mallard 76
Martin,
 Caribbean 358
 Cuban 248
 Purple 247
 Sand 251
Meadowlark, Eastern 332
Merganser,
 Hooded 85
 Red-breasted 86
Merlin 105
Mockingbird,
 Bahama 269
 Northern 268
Munia, Chestnut 368
 Scaly-breasted 369
 Tricoloured (Chestnut) 370

N

Night-heron,
 Black-crowned 63
 Yellow-crowned 64
Nighthawk,
 Antillean 200
 Common 199
Nightjar, Cuban 202
Noddy, Brown 167

O

Oriole,
 Baltimore 341
 Cuban 337–9
 Hooded 366
 Orchard 340
Osprey 89
Ovenbird 295
Owl,
 Bare-legged 192–3
 Barn 191
 Burrowing 196
 Long-eared 356
 Short-eared 198
 Stygian 197
Oystercatcher, American 126

P

Parakeet, Cuban 183–4
Parrot, Cuban 185
Parula, Northern 275
Pelican, American White 47
Petrel, Black-capped 44
Pewee, Cuban 225
Phalarope,
 Grey (Red) 352
 Red-necked 352
 Wilson's 352
Pheasant, Ring-necked 367
Phoebe, Eastern 227
Pigeon,
 Plain 171
 Rock 368
 Scaly-naped 169
 White-crowned 170
Pintail,
 Northern 80
 White-cheeked 79
Pipit, American 361
Plover,
 American Golden 120
 Black-bellied (Grey) 119
 Piping 124
 Semipalmated 123
 Snowy 121
 Wilson's 122
Potoo, Northern 203
Pygmy-Owl, Cuban 194–5

Q

Quail-dove,
 Blue-headed 181–2
 Grey-fronted 178–9
 Key West 177
 Ruddy 180

R

Rail,
 Black 108
 Clapper 109
 King 110
 Spotted 113
 Virginia 349
 Zapata 112, 371
Redhead 62
Redstart, American 292
Robin, American 265
Ruff 351

S

Sanderling 138
Sandpiper,
 Buff-breasted 351
 Least 141
 Pectoral 143
 Semipalmated 139
 Solitary 132
 Spotted 134
 Stilt 145
 Upland 135
 Western 140
 White-rumped 142
Sapsucker, Yellow-bellied 217
Scaup, Lesser 84
Scoter, Surf 348
Shearwater,
 Audubon's 44
 Cory's 342
 Great 342
 Sooty 343
Shoveler, Northern 78
Skimmer, Black 168
Skua,
 Arctic 149
 Long-tailed 353
 Pomarine 149
 South Polar 353
Snipe, Wilson's 148
Solitaire, Cuban 260–1
Sora 111
Sparrow,
 Chipping 319
 Clay-coloured 319
 Grasshopper 321
 House 36
 Lark 320
 Lincoln's 321
 Rufous-collared 364
 Savannah 320
 White-crowned 322
 Zapata 317–18
Spindalis, Western 309
Spoonbill, Roseate 67
Starling 360
Stilt, Black-necked 127
Stork, Wood 68

Storm-petrel,
    Band-rumped 344
    Leach's 343
    Wilson's 343
Swallow,
    Bahama 250
    Bank 251
    Barn 253
    Cave 252
    Cliff 251
    Northern Rough-winged 250
    Tree 249
Swan, Tundra 347
Swift,
    Black 204
    Chimney 206
    White-collared 205

T

Tanager,
    Scarlet 308
    Summer 307
Teal,
    Blue-winged 77
    Cinnamon 348
    Green-winged 81
Tern,
    Arctic 355
    Black 166
    Bridled 164
    Caspian 157
    Common 161
    Forster's 162
    Gull-billed 156
    Large-billed 355
    Least 163
    Roseate 160
    Royal 158
    Sandwich 159

Sooty 163
Thrasher, Brown 360
Thrush,
    Bicknell's 262
    Grey-cheeked 263
    Hermit 360
    Red-legged 266
    Swainson's 264
    Wood 265
Tody, Cuban 213–14
Towhee, Green-tailed 364
Trogon, Cuban 211–12
Tropicbird,
    Red-billed 344
    White-tailed 45
Turnstone, Ruddy 136

V

Veery 262
Vireo,
    Black-whiskered 243
    Blue-headed 241
    Cuban 238–9
    Philadelphia 242
    Red-eyed 244
    Thick-billed 237
    Warbling 241
    White-eyed 236
    Yellow-throated 240
Vulture,
    Black 91
    Turkey 90

W

Warbler,
    Bay-breasted 288

Black-and-white 291
Black-throated Blue 280
Black-throated Green 282
Black-throated Grey 361
Blackburnian 283
Blackpoll 289
Blue-winged 271
Canada 305
Cape May 279
Cerulean 290
Chestnut-sided 277
Connecticut 363
Golden-winged 272
Hooded 304
Kentucky 298
Kirtland's 326
Magnolia 278
Mourning 363
Nashville 274
Olive-capped 285
Orange-crowned 361
Oriente 302–3
Palm 287
Pine 362
Prairie 286
Prothonotary 293
Swainson's 295
Tennessee 273
Townsend's 362
Wilson's 305
Worm-eating 298
Yellow 276

Yellow-headed 300–1
Yellow-rumped 281
Yellow-throated 284
Waterthrush,
    Louisiana 297
    Northern 296
Waxwing, Cedar 270
Wheatear, Northern 369
Whip-poor-will,
    Eastern 356
Whistling-duck,
    Black-bellied 70
    Fulvous 72
    West Indian 71
    White-faced 346
Wigeon, American 75
Willet 133
Wood-pewee,
    Eastern 224
    Western 223
Woodpecker,
    Cuban Green 218–19
    Ivory-billed 223, 371
    West Indian 216
Wren,
    House 359
    Zapata 254–5

Y

Yellowlegs,
    Greater 130
    Lesser 131
Yellowthroat,
    Common 299

INDEX OF SCIENTIFIC NAMES

A
Accipiter
   gundlachi 96–7
   striatus 98
Actitis macularius 134
Agelaius
   assimilis 329–30
   humeralis 331
Aix sponsa 73
Ajaia ajaja 67
Alle alle 355
Amazona leucocephala 185
Ammodramus
   savannarum 321
Anas
   acuta 80
   bahamensis 79
   crecca 81
   fulvigula 348
   platyrhynchos 76
Anhinga anhinga 51
Anous stolidus 167
Anser
   albifrons 346
   caerulescens 347
Anthus rubescens 361
Antigone canadensis 118
Antrostomus
   carolinensis 201
   cubanensis 202
   vociferus 356
Aramus guarana 117
Archilocus colubris 208
Ardea
   alba 56
   herodias 55
Ardenna
   gravis 342
   grisea 343
Arenaria interpres 136
Asio
   flammeus 198
   otus 356
   stygius 197
Athene cunicularia 196
Aythya
   affinis 84
   americana 82
   collaris 83
   valisineria 82

B
Bartramia longicauda 135
Bombycilla cedrorum 270
Botaurus lentiginosus 53
Branta canadensis 347
Bubulcus ibis 61
Bucephala albeola 349
Buteo
   jamaicensis 102
   platypterus 100–1
   swainsoni 349
Buteogallus gundlachii 99
Butorides virescens 62

C
Cairina moschata 367
Calcarius lapponicus 364–5
Calidris
   alba 138
   alpina 144
   canutus 137
   fuscicollis 142
   himantopus 145
   mauri 140
   melanotos 143
   minutilla 141
   pugnax 351
   pusilla 139
   subruficollis 351
Calonectris borealis 342
Campephilus
   principalis 223
Caracara cheriway 103
Cardellina
   canadensis 305
   pusilla 305
Cathartes aura 90
Catharus
   bicknelli 262
   fuscescens 262
   guttatus 360
   minimus 263
   ustulatus 264
Chaetura pelagica 206
Charadrius
   melodus 124
   nivosus 121
   semipalmatus 123
   vociferus 125
   wilsonia 122
Chlidonias niger 166
Chlorostilbon ricordii 207
Chondestes
   grammacus 320
Chondrohierax wilsonii 91
Chordeiles
   gundlachii 200
   minor 199
Chroicocephalus
   philadelphia 151
   ridibundus 354
Circus hudsonius 95
Coccyzus
   americanus 187
   erythropthalmus 186
   merlini 189
   minor 188
Coereba flaveola 306
Colaptes
   auratus 220
   fernandinae 221–2
Colinus virginianus 107
Columba livia 368
Columbina passerina 176
Contopus
   caribaeus 225
   sordidulus 223
   virens 224
Coragyps atratus 91
Corvus
   minutus 244–5
   nasicus 246
   splendens 358
Crotophaga ani 190
Cyanerpes cyaneus 310
Cyanolimnas cerverai 112
Cygnus columbianus 347
Cypseloides niger 204

D
Dendrocygna
   arborea 71
   autumnalis 70
   bicolor 72
   viduata 346
Dolichonyx oryzivorus 328
Dumetella carolinensis 267

E
Egretta
   caerulea 58
   rufescens 60
   thula 57
   tricolor 59
Elanoides forficatus 92
Empidonax
   alnorum 357
   flaviventris 226
   traillii 227
   virescens 226
Eudocimus ruber 345

F
Falco
   columbarius 105
   peregrinus 106
   sparverius 104
Ferminia cerverai 254–5
Fregata magnificens 52
Fulica americana 116

G
Gallinago delicata 148
Gavia immer 342

Gelochelidon nilotica 156
Geothlypis
　formosa 298
　philadelphia 363
　trichas 299
Geotrygon
　caniceps 178–9
　chrysia 177
　montana 180
Glaucidium siju 194–5

H
Haematopus palliatus 126
Haliaeetus
　leucocephalus 94
Hapalocrex flaviventer 112
Helmintheros
　vermivorus 294
Himantopus
　mexicanus 127
Hirundo rustica 253
Hydrobates castro 344
Hydroprogne caspia 157
Hylocichla mustelina 265

I
Icteria virens 306
Icterus
　cucullatus 366
　galbula 341
　melanopsis 337–9
　spurius 340
Ictinia mississippiensis 92
Ixobrychus minutus 54

J
Jacana spinosa 129

L
Larus
　argentatus 153
　delawarensis 152
　fuscus 154
　marinus 155
Laterallus jamaicensis 108
Leiothlypis
　celata 361
　peregrina 273
　ruficapilla 274
Leucophaeus atricilla 150
Limnodromus
　griseus 146
　scolopaceus 147
Limnothlypis
　swainsonii 295
Limosa
　fedoa 350
　haemastica 350
Lonchura
　atricapilla 368
　malacca 370
　punctulata 369
Lophodytes cucullatus 85

M
Mareca
　americana 75
　strepera 74
Margarobyas lawrencii 192–3
Megaceryle alcyon 215
Melanerpes
　superciliaris 216
Melanitta perspicillata 348
Mellisuga helenae 209–10
Melopyrrha nigra 311–12
Melospiza lincolnii 321
Mergus serrator 86
Mimus
　gundlachii 269
　polyglottus 268
Mniotilta varia 291
Molothrus
　ater 366
　bonariensis 336
Morus bassanus 345
Myadestes elisabeth 260–1
Mycteria americana 68
Myiarchus
　crinitus 228
　sagrae 229

N
Nomonyx dominicus 87
Numenius americanus 350
Numida meleagris 368
Nyctanassa violacea 64
Nyctibius jamaicensis 203
Nycticorax nycticorax 63

O
Oceanites oceanicus 343
Oceanodroma
　leucorhoa 343
Oenanthe oenanthe 359
Onychoprion
　anaethetus 164
　fuscatus 163
Oporornis agilis 363
Oxyura jamaicensis 88

P
Pandion haliaetus 89
Pardirallus maculatus 113
Parkesia
　motacilla 297
　noveboracensis 296
Passer domesticus 369
Passerculus
　sandwichensis 320
Passerina
　amoena 365
　caerulea 324
　ciris 326
　cyanea 325
Patagioenas
　inornata 171
　leucocephala 170
　squamosa 169
Pelecanus
　erythrorhynchos 47
Petrochelidon
　fulva 252
　pyrrhonotoa 251
Phaethon
　aethereus 344
　lepturus 45
Phaetusa simplex 355
Phalacrocorax
　auritus 50
　brasilianus 49
Phalaropus
　fulicarius 352
　lobatus 352
　tricolor 352
Phasianus colchicus 367
Pheucticus
　ludovicianus 323
　melanocephalus 365
Phoenicopterus ruber 69
Pipilo chlorurus 364
Piranga
　olivacea 308
　rubra 307
Plegadis
　chichi 346
　falcinellus 66
Pluvialis
　dominica 120
　squatarola 119
Podilymbus podiceps 42
Polioptila
　caerulea 257
　lembeyei 258–9
Porphyrio martinicus 114
Porzana carolina 111
Priotelus temnurus 211–12
Progne
　cryptoleuca 248

dominicensis 358
subis 247
Protonotaria citrea 293
Psittacara euops 183–4
Pterodroma hasitata 44
Ptiloxena atroviolacea 333–4
Puffinus iherminieri 44
Pyrocephalus rubinus 356

Q
Quiscalus niger 335

R
Rallus
  crepitans 109
  elegans 110
  limicola 349
Recurvirostra
  americana 128
Regulus calendula 256
Rhynchops niger 168
Riparia riparia 251
Rissa tridactyla 354–5
Rostrhamus sociabilis 93

S
Sayornis phoebe 227
Seiurus aurocapillus 295
Setophaga
  americana 275
  caerulescens 280
  castanea 288
  cerulea 290
  citrina 304
  coronata 281
  discolor 286
  dominica 284
  fusca 283
  kirtlandii 362
  magnolia 278
  nigrescens 361
  palmarum 287
  pensylvanica 277
  petechia 276
  pinus 362
  pityophila 285
  ruticilla 292
  striata 289
  tigrina 279
  townsendi 362
  virens 282
Sialia sialis 359
Sicalis flaveola 363
Spatula
  clypeata 78
  cyanoptera 348
  discors 77
Sphyrapicus varius 217
Spindalis zena 309
Spiza americana 327
Spizella
  pallida 319
  passerina 319
Starnoenas
  cyanocephala 181–2
Stelgidopteryx
  serripennis 250
Stercorarius
  longicaudus 353
  maccormicki 353
  parasiticus 149
  pomarinus 149
Sterna
  dougallii 160
  forsteri 162
  hirundo 161
  paradisaea 355
Sternula antillarum 163
Streptopelia decaocto 172
Streptoprocne zonaris 205
Sturnella magna 332
Sturnus vulgaris 360
Sula
  dactylatra 344
  leucogaster 46
  sula 345

T
Tachybaptus
  dominicus 42
Tachycineta
  bicolor 249
  cyaneoviridis 250
Teretistris
  fernandinae 300–1
  fornsi 302–3
Thalasseus
  maximus 158
  sandvicensis 159
Tiaris
  bicolor 316
  canora 313–14
  olivacea 315
Todus multicolor 213–14
Torreornis inexpectata 317–18
Toxostoma rufum 360
Tringa
  flavipes 131
  melanoleuca 130
  semipalmata 133
  solitaria 132
Troglodytes aedon 359
Turdus
  migratorius 265
  plumbeus 266
Tyrannus
  caudifasciatus 234
  cubensis 232–3
  dominicensis 231
  forficata 235
  melancholicus 357
  savana 358
  tyrannus 230
  vociferans 357
Tyto alba 191

V
Vermivora
  chrysoptera 272
  cyanoptera 271
Vireo
  altiloquus 243
  crassirostris 237
  flavifrons 240
  gilvus 241
  griseus 236
  gundlachii 238–9
  olivaceus 244
  philadelphicus 242
  solitarius 241

X
Xanthocephalus
  xanthocephalus 366
Xema sabini 354
Xiphidiopicus
  percussus 218–19

Z
Zenaida
  asiatica 173
  aurita 174
  macroura 175
Zonotrichia
  capensis 364
  leucophrys 322

QUICK INDEX TO THE MAIN GROUPS OF BIRDS IN THIS BOOK

Anhinga	51
Bananaquit	306
Barn Owl	191
Boobies	46
Bullfinches, Grassquits and Sparrows	311
Cardinals	307
Cormorants	49
Crows	244
Cuckoos	186
Ducks	70
Falcons	103
Flamingos	69
Frigatebirds	52
Gnatcatchers	257
Grebes	43
Grosbeaks and Buntings	323
Gulls	150
Hawks and Eagles	91
Herons and Egrets	53
Hummingbirds	207
Ibises	65
Jacana	129
Jaegers	149
Kingfishers	215
Kinglets	256
Limpkin	117
Mockingbirds	268
New World Blackbirds and allies	329
New World Vultures	90
New World Warblers	271
Nightjars	199
Osprey	89
Oystercatchers	126
Parrots and Parakeets	183
Pelicans	47
Petrels	44
Pigeons and Doves	169
Plovers	119
Potoos	203
Rails	107
Sandhill Crane	118
Sandpipers and allies	130
Shearwaters	44
Skimmers	168
Solitaires	260
Spindalis	309
Spoonbill	67
Stilts and Avocets	127
Swallows	247
Swifts	204
Tanagers	310
Terns	156
Thrushes	262
Todies	213
Trogons	211
Tropicbirds	45
Typical Owls	192
Tyrant Flycatchers	223
Vireos	236
Waxwings	270
Wood Storks	68
Woodpeckers	216
Wrens	254